*Energy Policy in
the Greenhouse*

Energy Policy in the Greenhouse

Florentin Krause
Principal Investigator
Lawrence Berkeley Laboratory
Berkeley, California

Wilfrid Bach
University of Münster
Münster, Germany

Jonathan Koomey
Lawrence Berkeley Laboratory
Berkeley, California

A Report of the International Project
for Sustainable Energy Paths (IPSEP)
El Cerrito, California

John Wiley & Sons, Inc.
NEW YORK / CHICHESTER / BRISBANE / TORONTO / SINGAPORE

Library of Congress Cataloging in Publication Data:

Krause, Florentin, 1950–
   Energy policy in the greenhouse / Florentin Krause, Wilfrid Bach,
Jonathan Koomey.
     p. cm.
   "A Wiley-Interscience publication."
   ISBN-0-471-55663-7 (paper)
   1. Greenhouse effect, Atmospheric. 2. Energy policy. I. Bach,
Wilfrid. II. Koomey, Jonathan. III. Title.
QC912.3.K69 1992
363.73'87—dc20   91-45338
                                               CIP

Printed and bound in the United States of America by Braun-Brumfield, Inc.

10 9 8 7 6 5 4 3 2 1

Contents

Preface

This project was started in 1987, during the energy policy hiatus
brought on by the collapse of world oil prices, and following the
watershed events of the discovery of major acid rain damage and the
nuclear accident at Chernobyl. At the time, the oil price collapse had
aborted desirable price-driven changes in the energy system before
they had come to completion, and had fostered complacency. At the
same time, developments on the environmental front were sending
opposite signals. The result was a shift from energy strategy and policy
discussions driven mostly by economic considerations to a new era in
which environmental issues are becoming the principal driving force.
This trend toward redefining energy strategy and policy as instruments
for environmental protection is now culminating in the rapidly growing international concern over global warming.

These developments call for a new balance between environmental
protection, energy security, and economic concerns in the provision of
energy services. The new paradigm from which this balance could be
found is perhaps best described as *environmental least cost planning*:
the harnessing of cost-minimizing economic efficiency within norma-
tively defined environmental bounds and deliberate regulatory inter-
ventions to remove market barriers. How would this new approach
affect the cost of energy services, and economic welfare at large? This
question is a focus of this study on energy policy and global warming.

The project work began with a preliminary assessment of fossil-fuel
saving energy technologies. Important least-cost technology options

and related policy issues were presented and discussed in a two-day seminar sponsored by the European Economic Community (EEC) and the European Environmental Bureau (EEB) in June 1988. The seminar was attended by more than 120 representatives from non-government organizations and several governments, and included the Dutch Minister of Environment and the EEC Commissioners for Energy and Research and Development as speakers. Preliminary results of the work contained in this book were first presented at the OECD/IEA expert seminar on greenhouse gas emission reductions in April of 1989 in Paris.

Subsequent events, notably the Toronto meeting in June 1988, the establishment of the Intergovernmental Panel on Climate Change (IPCC) in the Fall of that year, and the preparations for the 1992 United Nations Conference on Environment and Development (UNCED), signaled the growing world-wide readiness to "think globally and act locally." At the same time, the official positions of a number of governments continued to be at odds both with the implications of the scientific evidence and with the need for political leadership on the global warming issue. Notably, the issue of emission reduction targets proved to be contentious. In response to the lack of consensus, a number of governments were ready to take unilateral measures that might build international consensus by example.

This conflict highlighted the need for independent scientific, economic, and political-logistic assessment of climate stabilization options. Furthermore, the development by individual countries of practical sectoral policies that could genuinely be called climate-stabilizing required the establishment of global, regional, and national greenhouse gas reduction targets. These targets, in turn, had to be more than numeric manipulations: they had to properly account for the issue of international equity and world social development in assigning future emission rights.

These considerations, and the lack of preexisting target analyses, led to the expansion of our project, which had originally been conceived more narrowly as a technical-economic examination of options to reduce carbon dioxide emissions in the energy sector of the industrialized countries. We decided to undertake the development of global and regional climate stabilization targets first, and to then develop energy scenarios that would meet these targets. Accordingly, the definition of energy policies as used in this book is considerably broader. The current book contains this most fundamental energy policy analysis, i.e., the analysis of climate stabilization targets. It starts from the science of the greenhouse effect, and uses climate modeling to arrive at fossil fuel consumption limits and timetables for their phaseout. The other part of our project, i.e., the assessment of the economic cost of reducing fossil carbon dioxide emissions through alternative energy strategies has been relegated to a forthcoming separate publication.

The current report was mainly drafted by Florentin Krause, who was the project leader and principal analyst. Wilfrid Bach contributed the climate modeling calculations and the compilations of greenhouse gas and other climate data. Jon Koomey helped execute many of the research tasks and was a valuable technical editor.

This volume has benefited from reviews and comments by a large number of people. The authors would like to acknowledge the contributions to the CFC data compilations and discussion by Lucas Reijnders; the carbon cycle model sensitivity calculations provided by Daniel Lashof based on a personal computer implementation by P. Moore; the time series of global and national gdp data provided by Kent Anderson; the provision of statistical data on global fossil carbon releases by Greg Marland; many computations, data manipulations, and graphic representations by Bruce Nordman; library reference assistance by Cheryl Wodley; and manuscript reviews by a number of analysts, including Daniel Lashof, Rob Swart, Walt Westman, Peter Hennicke, Bill Chandler, Ralph Cavanagh, and Ralph Torrie.

The design and production of the original report was ably executed by Byron Brown. Finally, the authors wish to thank Jacob Swager of the Dutch Ministry of Housing, Physical Planning and Environment, without whose enlightened, steadfast and generous support this project would have been impossible.

FLORENTIN KRAUSE
WILFRID BACH
JON KOOMEY

List of Tables

List of Figures

*Energy Policy in
the Greenhouse*

Part One

Climate Stabilization and the Limits to Fossil Fuel Consumption

Part One provides the scientific basis and risk-minimization arguments for specific limits on cumulative global fossil carbon emissions. As a core concept, a budget is developed for future carbon dioxide releases that would meet specified limits on the risks of human-induced climate change.

Chapter 1 reviews the current scientific understanding of the greenhouse effect, the history of the earth's climate, and the expected impacts from global warming. These data are used to define a plausible risk-minimizing warming limit. A methodology is defined for translating this limit into targets for practical policymaking.

Chapter 2 presents climate modeling calculations for emission scenarios of the major greenhouse gases. These calculations indicate what range of emission reductions would be needed to limit the risks of climate change to specified levels.

Chapter 3 analyzes reduction potentials for trace gases other than fossil carbon dioxide. We review the range of emission reductions that could be achieved on the basis of various technological options and policy measures.

Chapter 4 derives an upper-limit cumulative global budget for future fossil carbon dioxide releases. This budget reflects the potentials identified in Chapter 3 for reducing other greenhouse gas emissions. It is then compared with fossil fuel reserves and resources, and with projected fossil fuel consumption in several published global energy scenarios.

1

A Target-Based, Least-Cost Approach to Climate Stabilization

A. INTRODUCTION

1. Wait-and-See versus Risk Minimization

Despite recent progress in putting the threat of global warming onto the international agenda (see Section B below), the debate over the greenhouse effect continues to be shaped by two diametrically opposed viewpoints. These views can be characterized as follows:

> *Don't act until you are certain or wait-and-see.* Analysts holding to this view believe that current scientific uncertainties are still too large to warrant costly preventative action. Instead, more research should be pursued to reduce scientific uncertainties.

> *Act now to minimize risks.* Those holding to this view believe that current uncertainty cuts both ways: if major warming should come true, inaction could have catastrophic consequences. Society should therefore pursue investments and policies now to minimize such risks.

The competition between these two viewpoints revolves around the following fundamental issues:

- What aspects of the global warming threat are scientifically established fact, if any, and which are not?

- How costly would prevention be compared to adaptation?
- Would there be winners and losers, or would the consequences of global warming be catastrophic for the world as a whole?
- Is there reason to believe that remaining uncertainties could be satisfactorily resolved in a time frame that would still allow preventative global action later?
- Could improved scientific modeling tools be able to distinguish between winners and losers reliably?

What Is Established Scientific Fact and What Is Uncertain? The scientific community is in complete agreement that the atmospheric greenhouse effect governs global temperature.[1] In fact, heat entrapment due to radiative forcing of gases is one of the oldest and most well-established experimental findings of modern science, going back some 150 years. Moreover, the greenhouse effect is the only basis on which the enormous differences in atmospheric temperatures and climate between planets like Mars, Venus, and Earth can be explained.

What is also certain is that the atmospheric concentrations of a number of greenhouse gases have been rising and are continuing to rise, as shown by ongoing measurements in a global network of monitoring stations (Wuebbles and Edmonds 1988). There is compelling evidence that these increases are attributable to human activities (Dickinson and Cicerone 1986).

Furthermore, data from trapped air samples in ice cores have shown that for the last 150,000 years, atmospheric carbon dioxide and methane concentrations have closely tracked the surface temperature changes brought on by glacial and interglacial periods.[2]

There is also virtually no debate in the scientific community that continuing rises in the atmospheric concentrations of carbon dioxide and other greenhouse gases will lead to global warming (CDAC 1983; Schneider 1989).

Finally, scientific research to date has firmly established the *risk* of catastrophic consequences from future climate warming. There is

---

[1] For a recent review of the science and policy issues surrounding the greenhouse effect, see Schneider (1989).

[2] A number of other studies suggested that the same correlation also holds true for much earlier periods in the earth's climate, for which ice-core data are not available, but when carbon dioxide concentrations appear to have been twice the preindustrial level and the planet was 3°C warmer (Budyko et al. 1987; Sundquist and Brocker 1985). See Schneider and Londer (1984) for a discussion of the uncertainties in these estimates.

ample physical evidence that past changes in the earth's surface temperature were related to major changes in sea levels, ice cover, forest cover, and regional climates. If these changes were to occur again in the world of today, their consequences would likely be catastrophic: imagine, for example, a global sea-level rise of several meters (see Section D below). Though they might not occur, similar outcomes cannot be excluded as a consequence of future global warming.

What is uncertain is *how much* warming the earth will experience for a given increase in greenhouse gas concentrations (i.e., the climate sensitivity). It is also uncertain what the precise global and regional magnitudes and kinds of impacts will be: whether impacts will arrive gradually or suddenly; whether they will be catastrophic everywhere or only in some regions, and if so, where; and what monetary costs and benefits would be associated with these impacts.

Could Research Resolve These Uncertainties in a Timely Manner? Research could certainly improve the modeling tools of scientists by extensively measuring the geophysical and biogeochemical processes that are involved in climate responses to greenhouse warming. A key area would be the development and validation of a fully coupled atmosphere–ocean model. But both the data-gathering and the computational tasks involved are so enormous that it would probably take one to two decades before results could be expected to improve modeling capabilities significantly (Schneider 1989).

Similarly, monitoring of climate change, though needed, will inherently be unable to answer the key question of climate sensitivity (i.e., the warming response to a doubled carbon dioxide concentration) in a timely manner. The emergence of *any* degree of warming from the statistical noise in the world's temperature record is estimated to be 5 to 25 years away. Once the phenomenon of warming has emerged from measurements, interpretations regarding climate sensitivity will still be impossible until the effect has become sufficiently large (see Chapter 2).

Meanwhile, because of the "memory" of the climate system, each year of monitoring and research without simultaneous preventative measures adds a further irrevocable increment to future warming. In fact, the sensitivity of the global climate to specific radiative forcings will *never* be resolved through measurement until the *full* climatic change is upon us.

In view of this difficulty of interpreting future temperature changes that could be obtained from global monitoring, Hansen (1989) argues that the soundest approach is to calculate the range of climate sensitivities that is consistent with past changes in global climate and atmospheric greenhouse gas concentrations—essentially data from warming experiments the earth conducted on itself. He reviews two sets of such paleoclimatic data: carbon dioxide and methane concentrations during the last 150,000 years as obtained recently from ice cores, and the surface temperature changes the earth underwent during the same period.

Though there is a strong correlation between the two sets of data, it is not clear whether the rises in CO_2 concentrations found in ice cores caused past warmings or whether past warmings brought on rising CO_2 concentrations. Hansen argues that this question is immaterial for the issue of climate sensitivity. In his view, the correlation between the two parameters describes how the greenhouse effect has governed the earth–atmosphere system in the past, irrespective of the sequence of events that set these changes in motion. His calculations result in approximately the same range of climate sensitivities as currently predicted by the major climate models (see Chapter 2).

These calculations thus add empirical weight to the climate sensitivities obtained from a bottom-up combination of scientific knowledge about individual geophysical processes, as done in climate modeling. Like recent satellite data on the water vapor effect (Raval and Ramanathan 1989), these paleoclimatic data reinforce the scientific basis for predictions of significant global warming. At the same time, they do not narrow the substantial range of warming effects obtained from current models.

If the exact model-based quantification of climate change will remain elusive on the global level, this will be even more the case for predictions on a regional level. Nations that choose adaptation over collective prevention (in hopes of climatic improvements in their region) will be playing Russian roulette.

The wait-and-see approach is equivalent to converting the natural environments of all peoples and species into one huge laboratory—or, since this experiment has already inadvertently gotten under way, to continue and expand it. Any scientific research council would oppose such an experiment as unethical and irresponsible if it were proposed as a research project.

Risk Minimization as a Basis for Greenhouse Policy. The wait-and-see policy ignores that incurring risks has its own cost. Costs are only seen as existing on the prevention side of the ledger, while the risk-reducing benefits of preventative action are discounted. The principal rationale, that is, that scientific uncertainty could be sufficiently resolved through research to eventually allow the application of conventional cost-benefit analysis, is faulty. The continued attractiveness of a wait-and-see policy among some constituencies is mainly explained by a lack of information about the nature of the problem, as well as by unbridled technological optimism, and in some cases, vested economic interests in the status quo.

The risk-minimization approach to global warming, on the other hand, relies on a properly scientific outlook—not just in terms of the facts and risks that science has already established beyond question, but also in recognizing the inherent limits in striving for scientific certainty or, for that matter, for comprehensive and reliable monetary assessments of potential impacts.

In this paradigm, risks contribute to real costs. These risk-based costs can be expressed in the following simple equation:

(low or uncertain probability) ✕
(catastrophic consequences) = (major risk to society)

This perspective on risks is by no means unique to global warming. Huge military outlays are routinely made on the basis of this formula. Given the magnitude of climate risks (see Section D), global warming —and other environmental threats as well—could be treated as a new type of threat to global security.[3] Just as military expenditures are justified as precautionary measures that buy insurance against perceived risks and threats, precautionary measures to reduce greenhouse gas emissions could be seen as a form of *buying insurance against the risk and threat of climate change.*

In one form or another, most existing environmental regulation is already based on this formula. In all cases, normative, risk-based perceptions have had to take over where scientific analysis and cost-benefit calculus reached their limits. The relevance of these public policy precedents for the greenhouse issue is perhaps best illustrated by toxic substance regulation, notably that of new substances that could be carcinogens in humans. Carcinogen regulation is of particular interest for climate stabilization policy because it deals with scientific dilemmas very similar to those posed by the greenhouse effect. It is worth pointing out these parallels:

---

[3]This approach has been suggested, among others, by the Bruntland report to the United Nations Commission on Environment and Development (Bruntland Commission 1988).

- Scientific uncertainties about the carcinogenicity of substances cannot be resolved directly, that is, through experiment on humans. Established public policy therefore regulates or bans chemicals even when there is no definite proof of their toxicity. It is sufficient that they are suspected to be toxic on the basis of indirect evidence. The goal of regulation actually is never to find out for sure whether these substances are indeed toxic to humans, because that certainty could only be obtained at the cost of willfully endangering human lives. A rational response to the threat of global warming would have the same goal of protecting humankind from direct empirical evidence about its actual consequences, since such evidence could only be obtained through subjecting the world environment to potentially catastrophic climate change.
- Toxic substance regulation uses, among other things, normative risk limits, such as a maximum number of additional expected deaths due to exposure to a certain chemical. A warming limit as used in our climate analysis (see Section E) is equivalent in function to this regulatory standard.
- Because experiments on humans would be unacceptable, toxic substance regulation is based on animal tests and other indirect methods for determining risks. In the greenhouse context, such a surrogate function is provided by the climate modeling experiments commonly used to calculate the degree of warming from a given emission trajectory.
- To provide an additional safety factor in transferring results from animal tests to human risks, toxicological tests are conducted at high dosages per kilogram of animal body weight. Prevention-oriented climate policies should provide similar built-in safety factors by relying on the higher climate sensitivity estimates and on those models of global chemical cycles that require higher emission reductions to stabilize concentrations (see also Section E and Chapter 4).

2. Climate Stabilization and Sustainable Development: A Least-Cost Approach

How could the chasm between the monetized cost-benefit perspective and the risk-based perspective be bridged? To see this, it is necessary to recognize a common concern that is shared by both sides in the debate: the economic cost of prevention. In the wait-and-see approach, the underlying concern is that prevention would be expensive, while the benefits are uncertain or unknowable.

The risk-minimization approach raises a similar issue: Society may not have the resources to hedge against all future environmental risks. How much precautionary investment is affordable before the cure proves worse than the ailment, or before the political willingness of the present generation to make economic sacrifices for future generations is exhausted?

Both approaches presume that precautionary action is likely to be economically burdensome or disruptive. But the risk-minimizing measures would involve such consequences is not at all established fact.

For one, the magnitude of prevention costs depends on the perspective from which they are viewed. We must ask: An economic burden for whom? Costs seen as a major burden by vested interests might be a minor burden from a societal perspective, or indeed a net benefit. In dealing with the concerns of vested interests, the issue may be less to manage a major absolute cost to society than to find workable incentive policies that could shift investments and consumption away from warming-related products while providing a sufficient transition time for status quo stakeholders.

But aside from finding the proper societal cost perspective and political compromises, climate stabilization could largely be achieved by global and national policy measures that should be implemented anyway—climate change or no climate change. We need to ask: What would be the net cost of buying insurance against the risks of climate warming if we were to emphasize greenhouse control measures that also have tangible societal benefits in other areas?

The societal "side" benefits of precautionary climate-stabilizing measures could be

- Economic benefits in areas where status quo arrangements are blocking economic efficiency, as in the end-use efficiency of energy consumption;
- Environmental benefits in areas such as urban air pollution, acid rain, and forest dieback; or
- Social benefits in areas where reforms aimed at meeting basic human needs would also dampen the release of greenhouse gases, as in the case of land reform and reforestation schemes in developing countries.

All that climate stabilization might mean is that policies aimed at these problems would be pursued more vigorously, causing possibly no more than a minor additional increment in costs and effort. This climate-related increment could be so small as to hardly be painful or disruptive.

Ironically, the growing failure of world economic development on other fronts (Brown et al. 1988) also increases the opportunities for cost-reducing synergistic climate stabilization.[4] The major error in the conventional debate over the affordability of warming prevention is to ignore that present development patterns are unsustainable *overall*, not just in the climate area. In assessing the cost of risk-minimizing measures, we need to take the goal of sustainable world development as our baseline. Accounting based on the status quo has become irrelevant.

Seen this way, the fundamental questions for any climate-stabilization analysis can be summarized as follows:

- What measures would be needed to return societal development to a sustainable path, before considering climate change?
- To what extent would climate-stabilizing action overlap with these corrective measures?
- What is the net economic and other cost of climate stabilization relative to pursuing these other measures that should be undertaken anyway?

These questions frame what we call a *least-cost approach to climate stabilization*. Since energy use currently is, and foreseeably will remain, the dominant driving force of global warming (see Chapter 2), we focus on least-cost options for reducing use of fossil fuels. This focus suggests itself because the necessary quantitative data for a least-cost analysis are far more developed in the case of energy than for other major sources of greenhouse gases.

Given that energy use has a dominant role in climate stabilization, the question arises as to whether one could identify a cost-effective energy strategy that would lower the unit cost of energy services to society below the status quo level while curtailing fossil fuels. If so, climate stabilization could conceivably be achieved *at negative net cost*. The energy sector could become a "golden goose" for climate stabilization, by generating savings that could pay, partially or in full, for the cost of those climate measures that do result in net costs to society.

---

[4]This approach has been suggested, among others, by the Bruntland report to the United Nations Commission on Environment and Development (Bruntland Commission 1987).

Indeed, earlier least-cost-oriented explorations of national[5] and global[6] energy futures strongly suggested such a possibility. Needless to say, such a finding, if broadly applicable, could have a major impact on the chances for an effective international climate convention. A major task in supporting such a process is thus the broad-based implementation of least-cost-oriented energy analyses in many regions and countries.

However, to make such analyses sufficiently specific and comparable, it is necessary to clarify what level of climatic risk minimization should be pursued. The kinds of fossil-fuel substituting measures required, and their net costs to society, strongly depend on the climate-stabilization targets we set for ourselves. This book therefore pursues the question of what climate-stabilization targets would be risk-minimizing, yet still practically feasible.

B. GROWING AWARENESS OF THE CLIMATE THREAT

The realization that the emission of trace gases might change the climate of the planet goes back more than 150 years. Fourier was probably the first to discuss the CO_2/greenhouse effect in 1827 by comparing it with the warming of air isolated under a glass plate (Bach 1982–1984). In 1896–1903, Arrhenius made the first climate change calculation. His result: A CO_2 increase by a factor of 2.5–3 would increase the global mean surface temperature by 8–9°C. In 1941, Flohn noted that man-made CO_2 production perturbs the carbon cycle, leading to a continual CO_2 increase in the atmosphere. In 1957, Revelle and Suess concluded that human activities were initiating a global geophysical experiment that would lead to detectable climatic changes in a few decades. In the same year, Keeling and co-workers started the first CO_2 measurement program on Mauna Loa (Hawaii) and at the South Pole as part of the International Geophysical Year.

In 1969 and 1970, a group of scientists undertook the first major studies on the climatic effects of human activities. The results were a "Study of Critical Environmental Problems" (SCEP 1970) and a "Study of Man's Impact on Climate" (SMIC 1971). Both served as input to the 1972 U.N. Conference on the Human Environment. As a reaction to the growing concern that human activities might alter climate, in 1979 the World Meteorological Organization convened a

[5]See, for example, Krause et al. (1980) and Olivier et al. (1983).
[6]See Lovins et al. (1981) and Goldemberg et al. (1988).

World Climate Conference in Geneva (WMO 1979). An important outcome of this meeting was an urgent appeal to the world's nations to

- Take full advantage of man's present knowledge of climate;
- Take steps to improve significantly that knowledge; and
- Foresee and prevent potential man-made changes in climate that might be adverse to the well-being of humanity.

Since then, a number of national programs and initiatives have emerged. For example, in preparation for the German Climate Program, a series of international conferences were held, sponsored by the Federal Environmental Agency of Germany. These were "Man's Impact on Climate" (Bach et al. 1979); "Interactions of Energy and Climate" (Bach et al. 1980); and "Food–Climate Interactions" (Bach et al. 1981). These meetings led to the formulation of a low-climate-risk energy strategy, which would

- Promote the more efficient end use of energy;
- Secure the expeditious development of energy sources that add little or no CO_2 to the atmosphere; and
- Keep global fossil fuel use, and hence CO_2 emissions, at the present level.

These conferences initiated a shift in emphasis from calls for more studies to calls for immediate policy action.

In the wake of these calls for action, many other countries started their own climate programs. In the United States, the Carbon Dioxide Assessment Committee of the National Academy of Sciences issued a major report (CDAC 1983), followed by a study on greenhouse gases by the U.S. Environmental Protection Agency (EPA) (Seidel and Keyes 1983), a major, five-volume, state-of-the-art summary of CO_2 research by the U.S. Department of Energy (U.S. DOE 1985), and a four-volume report on the effects of changes in ozone and climate by the EPA and UNEP (United Nations Environment Program) (Titus 1986). Most recently, the EPA published an analysis of climate-stabilization options (Lashof and Tirpak 1989).

Canada has established a CO_2 and climate program (Hengeveld 1987). In the Netherlands, an assessment of the CO_2 problems was undertaken by the Committee of the Health Council of the Netherlands (CHCN 1983). Various activities on trace gases and climate were initiated within the European Economic Community and the World Meteorological Organization (WMO 1985).

In 1986, upon the recommendation of the 1985 Villach meeting, an Advisory Group on Greenhouse Gases (AGGG) was set up jointly by ICSU (International Council of Scientific Unions), WMO (World Meteorological Organization) and UNEP (United Nations Environment Program). Its tasks are to review global and regional studies on greenhouse gases and to evaluate the rates of greenhouse gas increases and their effects.

The annually recurring ozone hole over Antarctica has led to an initiative among 17 European research centers to set up EUROPICA (European Program in Chemistry of the Atmosphere). While ozone will be the focal point of this scientific mission, many of the climate-related trace gases will also be studied.

In 1986, ICSU launched a decade-long International Geosphere-Biosphere Program (IGBP). Its objectives are

> to describe and understand the interactive physical, chemical and biological processes that regulate the total Earth system, the unique environment it provides for life, the changes that are occurring in that system, and the manner by which these changes are influenced by human actions.

A second World Climate Conference was held in 1990. In preparation for this meeting, the following conferences have been held:

- UNEP/WMO/ICSU: "Assessment Conference on the Role of Carbon Dioxide and Radiatively-Active Constituents in Climate Variation and Associated Impacts," Villach, Austria, October 1985.
- UNEP/WMO/ICSU: "The Effects of Future Climatic Changes on the World's Bioclimatic Regions and their Management Implications—A New Technical Agenda," Villach, Austria, September/October 1987.
- Government of the Netherlands/EC/WMO: "Interrelated Bioclimatic Land Use Changes," Noordwijkerhout, the Netherlands, October 1987.
- UNEP/RBF/GMF, etc.: "Priorities for Future Management—A New Policy Agenda," Bellagio, Italy, November 1987.
- EEC/government of the Netherlands/EEB/IPSEP: "Energy Policy and Climate Change: What Can Western Europe Do?" Brussels, June 1988.
- Ministry of the Environment of Canada/UNEP/WMO: "World

Conference on the Changing Atmosphere: Implications for Global Security," Toronto, June 1988.

- UNCSTD/UNEP/WMO: "Climate Change and Variability, and Social, Economic and Technological Development," Hamburg, November 1988.
- First Meeting of the Intergovernmental Panel on Climate Change (IPCC) under the sponsorship of UNEP and WMO, Geneva, November 1988.

The panel in the last conference listing consists of three working groups: one on modeling, one on impact studies, and one on response strategies. The purpose of the IPCC is to prepare the Second World Climate Conference in 1990 and to work toward a global climate convention (see Chapter 7).

The establishment of the IPCC is symbolic of a major broadening of the international climate warming agenda from basic research to preventative policy action. Several other milestones in this process should be mentioned:

- In 1987, the U.S. Congress passed a Global Climate Protection Act that ordered the EPA, the Office of Technology Assessment, and the Department of Energy to prepare reports on specific policy actions that could limit global warming.
- In the same year, the FRG established a special standing parliamentary commission, the Enquete Commission on "Precautionary Measures for the Protection of the Earth's Atmosphere." Its express mandate is to identify and evaluate specific measures that can reduce the emission of trace gases that endanger the world's climate and ozone layer. A final report is planned for 1990.
- In June 1988, the Toronto World Conference on the Changing Atmosphere called for a 20 percent reduction of fossil carbon dioxide releases by 2005. Two heads of state appealed for a global convention to deal with the greenhouse effect.
- In the wake of the Toronto meeting, several far-reaching legislative proposals were introduced in the U.S. Congress that called for reductions in carbon dioxide emissions in addition to reductions in chlorofluorocarbon emissions.
- In the U.S.–Soviet summit held at the end of 1988, greenhouse warming was recognized for the first time as an issue at that level.
- In February 1989, the government of Canada sponsored an "International Meeting of Legal and Policy Experts, Protection of the

Atmosphere," in Ottawa, with the purpose of examining formulations and principles for a global climate convention.

- In April 1989, the OECD's International Energy Agency held the first major international expert seminar on energy technologies for reducing emissions of greenhouse gases, attended by more than 200 participants from 24 nations.
- At the IPCC meeting in Geneva in May 1989, more than 20 major participants supported a global climate convention.
- The Helsinki declaration of May 1989 brought a major increase of the numbers of governments supporting action on the ozone and greenhouse problems.
- Around the same time, the Dutch government adopted a target to return fossil carbon emissions to the 1990 level by 2000, and the Swedish parliament voted to limit fossil carbon releases to present levels.
- During the Group of Seven Summit in July 1989 in Paris, global warming became an agenda item.

Events are proceeding so rapidly that the above chronology will soon be hopelessly out of date. Underlying this rising tide of international activity is the growing recognition of the enormous risks entailed in climate warming.

In summary, the greenhouse effect has been known as a scientific possibility for more than a century. But only in the last two decades has this threat begun to be taken seriously, and only during the last two years have preventative measures entered the international political arena. Most scientific analysts now agree that the only way to address the greenhouse threat is to reduce the emissions that are the driving forces of global warming.

C. DRIVING FORCES OF CLIMATE CHANGE

The causes of climatic change are complex, and there are many theories and possible mechanisms. We limit our discussion to the driving forces created by trace gases related to human activity, since those are the only ones we can influence through policy action.

1. The Greenhouse Effect

Like a windowpane in a greenhouse, a number of gases in the earth's atmosphere let solar radiation (visible light) pass to the surface of the earth while trapping infrared (IR) radiation, also known as heat radiation, that is reemitted by the surface of the earth. This heat radiation would have otherwise escaped to space. It is this trapping of infrared radiation that is referred to as the greenhouse effect.[7] Gases that influence the surface–atmosphere radiation balance are also called radiatively active or greenhouse gases (GHG).

Even in the absence of human interference, the greenhouse effect is constantly in operation in maintaining the earth's climate. A number of natural constituents of the atmosphere are radiatively active. The most important are water vapor, carbon dioxide (CO_2), and clouds. These contribute roughly 90 percent to the natural greenhouse effect, whereas naturally occurring ozone (O_3), methane (CH_4), and other gases account for the remainder.

Human activities cause the emission of a number of greenhouse gases. These emissions create a change in the radiative balance of the surface–atmosphere system (radiative forcing). Because the concentrations of natural and anthropogenic greenhouse gases are small compared to the principal atmospheric constituents of oxygen and nitrogen, these gases are also called trace gases.

2. Important Trace Gases

Trace gases fall into three categories: (1) *radiatively active trace gases*, such as water vapor (H_2O), carbon dioxide (CO_2), ozone (O_3), methane (CH_4), nitrous oxide (N_2O), and the chlorofluorocarbons (CFCs), which exert direct climatic effects; (2) *chemically/photochemically active trace gases*, such as carbon monoxide (CO), nitrogen oxides (NO_x), and sulfur dioxide (SO_2), which exert indirect climatic effects through the chemistry that determines the atmospheric concentrations of hydroxyl radicals (OH), CH_4, and O_3; and (3) *aerosol emissions*.

Table 1.1 lists most greenhouse gases. Some 40 such gases have been identified so far, most of which are radiatively active. The characteristics of the major greenhouse gases are summarized in Tables 1.2a–c. These tables show current concentrations in the atmosphere, current concentration trends, atmospheric residence times, direct climate and chemical effects, and chemical–climate interactions. Detailed global

---

[7]At a global average surface temperature of 288°K or 15°C, the long-wave outgoing radiation from the surface of the earth is 390 W/m², compared to 236 W/m² from the top layer of the atmosphere. This reduction in the long-wave emission is a measure of the greenhouse effect (Ramanathan et al. 1987).

TABLE 1.1   Overview of Greenhouse Gases in the Atmosphere

TRACE GAS	CHEMICAL SYMBOL	TRACE GAS	CHEMICAL SYMBOL
Carbon Group		*Halogen group*	
Carbon monoxide	CO	Trichlorofluoromethane	$CFC1_3$
Carbon dioxide	CO_2	(Freon 11)	
Methane	CH_4	Dichlorodifluoromethane	CF_2C1_2
Oyxgen group		(Freon 12)	
Ozone	O_3	Chlorotrifluoromethane	CF_3Cl
		(Freon 13)	
Nitrogen group		Dichlorofluoromethane	$CFHC1_2$
Nitrous oxide	N_2O	(Freon 21)	
Nitrogen dioxide	NO_2	Chlorodifluoromethane	CF_2HCl
Dinitrogen pentoxide	N_2O_5	(Freon 22)	
Nitric acid	HNO_3	Trichlorofluoroethane	CF_3CC1_3
Ammonia	NH_3	(Freon 113)	
Sulfur group		Dichlorotetrafluoroethane	$C_2F_4C1_2$
Sulfur dioxide	SO_2	(Freon 114)	
Sulfur hexafluoride	SF_6	Chloropentafluoroethane	C_2F_5Cl
Carbonyl sulfide	COS	(Freon 115)	
Carbon disulfide	CS_2	Hexafluoroethane	C_2F_6
Hydrogen sulfide	H_2S	(Freon 116)	
Non-methane hydrocarbons		Methyl bromide	CH_3Br
Acetylene	C_2H_2	Ethylene bromide	$BrCH_2CH_2Br$
Acetaldehyde	CH_3CHO	Bromotrifluoromethane	CF_3Br
Formaldehyde	HCHO	Methyl chloride	CH_3Cl
Ethylene	C_2H_4	Methylene chloride	CH_2C1_2
Ethane	C_2H_6	Dichloroethane	CH_2ClCH_2Cl
Propane	C_3H_8	Trichloroethylene	C_2HC1_3
Butane	C_4H_{10}	Tetrachloroethylene	C_2C1_4
Methyl pentane	C_4H_{14}	Methyl chloroform	CH_3CC1_3
Others		Carbon tetrafluoride	CF_4
Peroxyacetyl nitrate	PAN	Carbontetrachloride	CCl_4

After WMO (1982), Chamberlain et al. (1982), and Ramanathan et al. (1985).

budgets for the sources and sinks of most of these gases can be found in Tables 3.2–3.7. A brief summary follows below.

Carbon Dioxide (CO_2). CO_2 is a radiatively active trace gas. Some 96 percent of the CO_2 is released by natural sources. The ocean and the biosphere are the main sinks. The remaining 4 percent stem from anthropogenic sources, mostly fossil fuel burning and land-use conversion (i.e., deforestation and soil degradation).

Methane (CH_4). CH_4 is a radiatively active trace gas. More than half of total emissions are from anthropogenic sources. The main sources are

TABLE 1.2a Overview of Key Characteristics of Trace Gases: CO_2, CH_4, and N_2O

Characteristics	Trace Gases		
	CO_2	CH_4	N_2O
1) Current conc. in atm.	1) 346 ppm (1986)	1) 1.7 PPM (1985)	1) 0.31 ppm (1985)
2) Atmospheric residence time	2) ca. 100 yrs. (for atmosphere, biosphere and upper ocean)	2) 7–10 yrs.	2) ca. 170 yrs.
3) Current conc. trend	3) ca. + 0.4 %/yr	3) 1.1 ± 0.1 %/yr (1951–1983)	3) 0.2–0.3 %/yr
4) Radiatively and chemically interactive	4) One of the most important infrared absorbers in ranges 550–800; 850–1100; 2100–2400 cm^{-1}	4) Infrared absorber (950–1650 cm^{-1}). Chemically reactive in both trop. and strat.	4) A strong infrared absorber (520–660; 1200–1350; 2120–2270 cm^{-1}). Inert in trop. Destroyed in strat. through photolysus.
5) Direct climate effect	5) The global mean equilibrium surface temp. increase due to the direct radiative effect of a CO_2 doubling (2 x CO_2) is $\Delta T_e = 1.3$ K; incl. climatic feedbacks it is $\Delta T_e = 1.5$–4.5 K.	5) For 2 x CH_4 $\Delta T_e =$ 0.2–0.4 K.	5) For 2 x N_2O $\Delta T_e =$ 0.3–0.4 K
6) Index of radiative sensitivity refers to expected growth from 1980–2050. The index for CO_2 is 10.	6) A relative measure of the contribution of a gas to radiative forcing. A surface temperature change $\Delta T_S = 2$ K for CO_2 gives a value of 10.	6) If $CO_2 = 10$ $CH_4 \approx 2 \pm 1$	6) If $CO_2 = 10$ $N_2O \approx 1.4 \pm 1$
7) Direct chemical effect	7) None	7) Increasing CH_4 • reduces OH in trop. • produces O_3 in trop. • produces O_3 in lower strat. • decreases O_3 in upper strat. through HO_x production • increases H_2O in strat. • decreases effectiveness of Cl and NO_x on O_3 in strat.	7) Through reaction of N_2O with $O(^1D)$ main source of NO_x in strat. Decreases O_3 in strat. Interacts with ClO_x and HO_x.
8) Chemical-climate interactions	8) Changes in the atmosphere's temperature profile affect reaction rates of gases. Temperature changes in strat. change ozone distribution which feeds back on surface climate.	8) Reduction of OH in trop. slows removal of CH_4 and other trace gases. Effects on O_3 and H_2O have climatic implications. Warmer climate changes biochemical sources of CH_4. Changes in temp. distribution affect atmospheric loss rate of CH_4.	8) O_3-destruction in strat. has climatic implications. Warming of trop. affects biogenic source rates.
9) Uncertainties in radiative transfer	9) Line half-widths and temperature dependencies and also overlap with H_2O need refinement.	9) Line half-widths and temp. dependence on CH_4 need to be better evaluated.	9) The spectral line intensities need further study.
10) Uncertainties in chemistry	10) None	10) Reactivity of CH_4 oxidation byproducts and their effects on O_3 in trop. need to be evaluated. The reaction of CH_4 with Cl in strat. is major uncertainty which could affect chemical impacts of CFCs on O_3.	10) Further study needed of • mechanisms and production rates from biogenic and combustion sources in trop. • the photodissociation rate of N_2O in strat.
11) Other	11) Further flux measurements needed to refine understanding of biogenic sources. Needed for making better projections of future. Studies of biomass growth due to CO_2-stimulation must take into account counteracting effects of acid rain and O_3, etc., (e.g., forest dieback).	11) More information is needed on the emission fluxes from various sources and the causes of the current CH_4 trends.	11) Better measurements of sources and sinks required. Biogenic source response to temperature change is largely unknown.
12) Climate effect of one molecule rel. to CO_2	1	25–32	150–250

Source: Wuebbles and Edmonds (1988), Lashof and Tirpak (1989), and UNEP (1987).

TABLE 1.2b Overview of Key Characteristics of Trace Gases: CFC-11 and CFC-12, O_3, and OH

Characteristics	Trace Gases			
	CFC-II	CFC-12*	O_3	OH
1) Current conc. in atm.	1) 0.20 ppb (1985) 0.32 ppb (1985)*		I) Trop.: 0.02–0.1 ppm; 2–3 times higher in urban areas Strat.: 0.1–10 ppm Based on ground-based and satellite data for 1978–1985.	I) In trop.: ca. 0.015 ppt In lower strat.: ca. 0.02 ppt In upper strat.: ca. 0.3 ppb.
2) Atmospheric residence time	2) ca. 65 yrs. ca. 110 yrs.*			2) Seconds to minutes.
3) Current conc. trend	3) ca. 5 %/yr		2) O_x (ozone + oxygen atoms) is	3) Largely unknown. Growth in CH_4 and CO suggests decrease in trop. by ca. 25% since 1900?
4) Radiatively and chemically interactive	4) Strong infrared absorbers (800–1200 cm^{-1}) (850–1250 cm^{-1})* inert in trop. Dissociation in strat. leads to highly reactive chlorine species.		· ca. 1 hr in upper strat. · months in lower strat. · hours to days in trop.	
5) Direct climate effect			3) Uncertain. But satellite and ground-based data indicate	4) Radiatively: no; chemically: yes.
6) Index of radiative sensitivity refers to expected growth from 1980–2050. The index for CO_2 is 10.	5) For a change from 0 to 2 ppb $\Delta T_s \approx 0.3$ K* 6) if $CO_2 = 10$ CFC-11 $\approx 2 \pm 1$ CFC-12 $\approx 4 \pm 1$*		· O_3 is decreasing in upper strat. · O_3 is increasing in trop. 4) A major absorber of both solar and infrared radiation (600–800, 950–1200 cm^{-1}). O_3 production and destruction strongly affected by chemical processes.	5) None. 6) None. 7) In trop.: · Controls lifetime and abundance of CH_4 and CO · Chief oxidant of CO, H_2O, CH_3CCl_3, NO_x and all HC · Initiates photochemical smog In strat.: · is a catalyst in O_3-destruction · affects other catalysts through reactions with Cl and NO_x.
7) Direct chemical effect	7) Dissociation leads to chlorine species which also react with HO_x and NO_x and together may lead to a significant O_3 destruction in strat.		5) Increase in trop. O_3 and decrease in surface in strat. O_3, increase surface temp. A 50% increase in trop. O_3 leads to a warming increase of ca. 0.3 K.	
8) Chemical-climate interactions	8) O_3 destruction in strat. has climatic implications in trop.		6) If $CO_2 = 10$ $O_3 \approx 2 \pm 1$	8) Has a strong indirect effect on climate due to its effect on radiatively important gases (CH_4, O_3, H_2O, CH_3CCl_3). It determines the ultimate effects of other trace species on O_3.
9) Uncertainties in radiative transfer	9) Especially the band strengths for infrared absorption		7) Trop. O_3 may change · directly due to CO, NO_x, CH_4, HC · indirectly due to strat. O_3 Strat. O_3 may change due to CFCs, N_2O, NO_x, H_2O, CH_4, CO_2 and circul.	
10) Uncertainties in chemistry	10) Especially the effects of odd-chlorine chemistry on O_3.		8) O_3 affects temp. distribution, which influences O_3 destroying mechanisms.	9) Radiative techniques to measure OH need further evaluation.
11) Other	11) Better emission data and concentration measurements both in trop. and strat. required also for other CFCs. Have high significance due to their direct effects on climate and their effects on the ozone layer.		10) Chemical reactions determining O_3 conc. Measurement techniques for more reliable trends. 11) Is one of the most important species · Plays major role in climate · In strat. protects us from UV-rad. · In trop. it is harmful to health and forests.	10) Rapid, reliable and accurate methods for measuring OH are needed. The reactivity of OH with HC requires more study. 11) Plays a significant role in climate and O_3 chemistry. Its global atmospheric abundance must therefore be better assessed.
12) Climate effect of one molecule rel. to CO_2	14000–17500	17000–20000	2000	NA

Source: Wuebbles and Edmonds (1988), Lashof and Tirpak (1989), and UNEP (1987).

TABLE 1.2c   Overview of Key Characteristics of Trace Gases: CO and NO_x

Characteristics	Trace Gases	
	CO	NO_x
1) Current conc. in atm.	I) NH: .20 ppm (1984) SH: 0.05 ppm (1984)	I) In trop. remote areas: 10–30 ppt In trop. populated areas: > 1000 ppt In strat.: ca. 0.02 ppm
2) Atmospheric residence time	2) 0.4 yrs (global); 0.1 yrs (in tropics)	2) In trop.: ca. 1–7 days; in strat.: much longer
3) Current conc. trend	3) 2 (1–5) %/yr	3) Very uncertain due to large spatial and temporal variations.
4) Radiatively and chemically interactive	4) is a weak absorber in the infrared. Reacts with OH, also produced in atmosphere by oxidation of CH_4 and other HC.	4) Radiatively yes, but not likely to be important. Chemically interactive through • reactions with OH, HO_2 in trop. • by determining O_3 distribution in strat.
5) Direct climate effect	5) No direct climate effect.	5) and 6) NO_2 is absorber of solar radiation but unlikely to be important. Indirect effects are not well known.
6) Index of radiative sensitivity refers to expected growth from 1980–2050. The index for CO_2 is 10.	6) No direct climate effect.	
7) Direct chemical effect	7) CO uses OH for oxidation which, in turn, affects the rate of oxidation of CH_4 and other HC. In the presence of NO_x it leads to O_3 formation in the trop. It reacts ultimately to form CO_2.	7) In trop.: NO and NO_2 are very active catalysts in reactions creating O_3 and photochemical smog. In strat.: NO_x are important chemical species determining O_3. NO_x also strongly influences reaction pathway for CH_4 oxidation.
8) Chemical-climate interactions	8) Affects climate through its effects on OH, O_3, CH_4 and other HC.	8) Its chemical interactions with other species changes O_3 and affects climate.
9) Uncertainties in radiative transfer	9) None.	9) Photodissociation of NO_x species (e.g., NO_3, N_2O_5) needs better understanding.
10) Uncertainties in chemistry	10) Increased CO may be linked to increased forest clearing and fossil fuel burning, increased CH_4 or decreased OH. Greater spatial and temporal variabilities make trend assessment more difficult than for CH_4 and CO_2. Need to develop airborne and satellite sampling techniques.	10) and 11) Global distribution not well understood. Uncertain NO_x reactions are: • reaction rates of odd-N with aerosols • reaction rates of nitric acid aerosols with chlorine species (important in explaining Antarctic ozone hole) • reactions involving organic nitrates in trop.
11) Other	11) As a toxic pollutant CO is controlled locally. As a trace gas affecting indirectly climate we need to assess its emissions and concentration time history.	

Source: Wuebbles and Edmonds (1988), Lashof and Tirpak (1989), and UNEP (1987).

cattle feedlots, biomass burning, rice-paddy fields, natural gas leakage, coal mining, and refuse disposal sites.

Nitrous Oxide (N_2O). N_2O is a radiatively active trace gas. The relative proportion of sources are uncertain. Cultivation of soils, soil fertilization, and biomass and fossil fuel burning are involved. The major sink is stratospheric photolysis with subsequent O_3 destruction.

Chlorofluorocarbons (CFC-11 and CFC-12). The CFCs have no natural sources; they are entirely man-made. They are radiatively active trace gases contributing to the tropospheric greenhouse effect. CFCs have no sinks in the troposphere. They are photodissociated in the stratosphere, where they are involved in the catalytic destruction of O_3.

Ozone (O_3). Ozone is a radiatively active trace gas that is formed as a product of natural and human-induced chemical and photochemical reactions in the lower and upper atmosphere (troposphere and stratosphere). Increases in tropospheric ozone are caused by emissions of common air pollutants such as carbon monoxide, nitrogen oxides, and hydrocarbons. At the same time, decreases in stratospheric ozone, which are caused by CFCs and some of the same pollutants, also lead to warming.

Hydroxyl Radicals (OH). This highly reactive compound is again the product of natural chemical processes in the atmosphere. It indirectly controls the abundance of several radiatively important trace gases, including ozone and methane. Conversely, anthropogenic emissions of chemically active trace gases such as carbon monoxide, hydrocarbons, and nitrogen oxides can reduce its abundance and thus increase radiatively active abundances.

Carbon Monoxide (CO). CO has no direct radiative effect, but through its reaction with OH radicals, it is exerting an indirect climatic effect by increasing concentrations of the radiatively active gases CH_4 and tropospheric O_3. Deforestation and fossil fuel burning are the two dominant man-made sources for CO.

Nitrogen Oxides (NO_x). Most NO_x emissions are of human origin. Biomass burning and fossil fuel combustion are the most important anthropogenic sources. In the troposphere, NO and NO_2 are active catalysts in reactions involving the radiatively active gases O_3 and CH_4. In the stratosphere, NO_x is important in determining O_3 levels.

Nonmethane Hydrocarbons (NMHC). Nonmethane hydrocarbon emissions from vegetation, vehicles, and industry are directly involved in urban smog and tropospheric ozone formation. They oxidize to carbon monoxide and thus also contribute to warming through the $OH - CH_4$ - tropospheric ozone link.

3. Trace Gas – Climate Interactions

The chemical constituents of the atmosphere can modify the climate through one or a combination of the following processes (Ramanathan et al. 1987; Wang et al. 1986; Wuebbles and Edmonds 1988).

Radiative Activity. Trace gases, which are radiatively active in the long-wave spectral region, will enhance the atmospheric greenhouse effect. The $7-13$-μm spectral region is referred to as the atmospheric "window" because some $70-90$ percent of the radiation emitted by the surface and clouds escapes to space. Most of the gases listed in Table 1.1 are strong absorbers in that region and are therefore quite effective in enhancing the greenhouse effect in the troposphere. Moreover, CO_2 and O_3 also have a significant influence on climate in the stratosphere by governing stratospheric long-wave emission and absorption.

Chemical Activity. Gases such as CO and NO, which have only minor radiative effects, are indirectly involved in changing the climate by altering the concentrations of radiatively active gases such as O_3 and CH_4 by chemical interactions (see Tables 1.2a–c).

Combined Radiative and Chemical Activity. This combined process occurs, for example, in the case of CH_4 and CFCs. Oxidation of CH_4 in the troposphere leads to more O_3, which, in turn, enhances the tropospheric greenhouse effect. The breakdown of CFCs in the stratosphere releases reactive chlorine, which destroys O_3. Depending on the vertical distribution of O_3 destruction, this can either amplify or dampen the surface warming due to CFCs (see Tables 1.2a and b).

Ozone Change and Stratosphere – Troposphere Radiative Interactions. Ozone is one of the most radiatively active gases because it absorbs solar radiation in addition to absorbing and emitting long-wave radiation. Tropospheric climate is influenced both by stratospheric O_3 through complex stratospheric–tropospheric radiative interactions and by the greenhouse effect of tropospheric O_3. All of this makes it difficult to state the nature, sign, or magnitude of the O_3 influence on climate (see Table 1.2b).

Radiative – Chemical interactions. The rates of the chain reactions that govern stratospheric O_3 are strongly temperature-dependent. The net effect is that a temperature decrease (increase) in the upper stratosphere results in an O_3 increase (decrease). For example, the CO_2 increase in the troposphere leads to tropospheric warming but also stratospheric cooling. This, in turn, leads to an increase in stratospheric O_3, which then compensates some of the CO_2-induced stratospheric cooling. An O_3 decrease, on the other hand, permits deeper penetration of solar radiation, leading to more O_3 production at lower stratospheric levels. This change in the O_3 profile has, in turn, an effect on the surface climate (see Table 1.2b).

Interactions Involving Stratospheric Dynamics. Climate change is strongly linked to large-scale motions in the stratosphere. Interactions between transport and chemistry determine the vertical and latitudinal distribution of O_3 change in the lower stratosphere, which, in turn, determines the climatic effect of O_3 change. There is also the effect of altered dynamics on the transport of trace gases resulting from changes in stratospheric adiabatic heating due to O_3 change. Finally, perhaps most important are the interactions between tropopause temperature, trace gas changes, and stratospheric H_2O.

Climate – Chemistry Interactions. The surface warming due to greenhouse gases increases evaporation, and hence, the H_2O content of the troposphere. Through photolysis and H_2O chemistry, this can perturb OH, the major cleansing and oxidizing agent for tropospheric gases. In addition, CO and NO_x react with OH to alter its concentration, which, in turn, changes the concentrations of such radiatively active gases as O_3 and CH_4 (see Table 1.2c).

4. Memory of the Climate System

The individual components of the climate system — such as the atmosphere, the hydrosphere, the cryosphere, the lithosphere, and the biosphere — have greatly varying response times. For example, an air molecule may remain in the troposphere for only 4–8 days, a water molecule in the deep ocean up to 1500 years, and an ice crystal in an ice sheet as long as ten thousand to a million years (Bach 1982–1984). This makes the climate system rather inert.

Of particular importance is the sluggishness in the warming of the ocean, which is the result both of the enormous heat capacity of the ocean and the long time constants involved in ocean circulation. Modeling studies of various degrees of sophistication show that the world's oceans can take from a decade to as much as a century to equilibrate with the changes in radiative heating induced by changes in atmospheric trace gas concentrations (Schlesinger 1986).[8]

The CO_2 uptake and release by the ocean is also a slow process, varying as a function of temperature as well as by atmospheric and oceanic C content. This is expressed as the fraction of anthropogenic net CO_2 emissions that remains in the atmosphere, that is, the "airborne fraction." Currently, this fraction is 50–60 percent.[9] Under equilibrium conditions in the ocean–atmosphere system, only about 20 percent of the emissions would remain airborne.

5. Feedback Mechanisms

Greenhouse warming has the potential to produce many complex feedbacks that are the major source of current climate-modeling uncertainties. These feedbacks could substantially increase the sensitivity of the climate system to human perturbation. They can be categorized as follows (Lashof 1989):

- Geophysical climate feedbacks (due to physical as opposed to chemical or biochemical processes); they include
  - Increased water vapor due to warmer and wetter climate.
  - Decreased reflectiveness (albedo) of the earth's surface due to shrinking snow and ice cover.
  - Increases in clouds due to greater evaporation.

- Biogeochemical feedbacks, including
  - Physical effects of warming (release of methane from hydrates in sediments; changes in ocean circulation and mixing affecting CO_2 uptake).
  - Climate–chemical feedbacks (changes in hydroxyl concentrations and tropospheric ozone due to more water vapor; shifts in the CO_2–carbonate equilibrium in the ocean).
  - Short-term biological responses to warming (increased microbial

---

[8]This response time τ_e is referred to as the "*e*-folding time," which implies a reduction of $\frac{1}{2.718}$ of the original value. If τ_e were close to ten years, the actual response of the climate system would be quite close to the equilibrium response; if, on the other hand, τ_e were close to 100 years, the actual response of the climate system would be quite far from the equilibrium response.

[9]In 1988, the airborne fraction of fossil carbon emissions increased to 90 percent. It is not clear whether this phenomenon is due to a loss of absorptive capacity of the oceans or some other cause (MacDonald 1989).

activity and therefore methane releases from soil organic matter; carbon dioxide fertilization of plant growth; increased plant respiration; increased nonmethane hydrocarbons from vegetation).

- Effects due to the reorganization of ecosystems (changes in surface albedo; changes in terrestrial carbon storage; changes in the biological pumping of carbon from the ocean surface to deep waters).

At this time, the biogeochemical feedbacks have not been incorporated in current climate models.

6. Implications for Climate Stabilization

Inertia and feedback mechanisms mean that the full warming impacts from greenhouse gas emissions manifest only with delay. The full warming impact from a change in greenhouse gas concentrations is also referred to as the *equilibrium warming*, while temperature effects on the way to reaching this equilibrium are called *transient warming*. If emissions continue at a sufficient rate to maintain raised levels of atmospheric concentrations, the transient warming will eventually converge on the equilibrium warming.

The full warming impact could be avoided if concentrations of trace gases in the atmosphere could be made to decline again before the impact of previous emissions has fully materialized. This possibility is important to recognize, though the degree to which such a reversal is possible is limited both by practical and physical factors. For the gases with long atmospheric residence times, just holding emission rates constant will lead to inexorable rises in atmospheric concentrations, and a complete cessation will only lead to slow declines in concentrations.

However, the short-lived greenhouse gases respond quickly to cuts in emissions, among them methane and ozone. Since both are related to air pollution sources, controlling the air pollution emissions that govern their tropospheric chemistry is one area in which short-term reductions in radiative forcing could be obtained.

D. GREENHOUSE RISKS AND CONSEQUENCES: JUST WARMER WEATHER?

1. Temperature History of the Earth's Climate

Several hundred million years ago, the earth may have been up to 20°C warmer than it is today, with no ice on both poles. In that time period, the planet moved through an ice age that produced glaciers in the inland areas of the tropical regions of Africa, India, and South America.

The more recent history of the earth's climate is revealed in data from tree rings, sea and lake levels, fossil pollen, ice cores, ice cover, altitudes and latitudes of tree lines, and cores from the ocean floor. These data sources allow approximate quantification of temperature conditions as far back as about 15–38 million years (the so-called *Oligocene/Miocene*). Based on these data, global average surface air temperature was about 6°C warmer than today (Bach et al. 1980).

Moving toward the present, the next warming peak occurred during the *Pliocene*, from about 3–5 million years before present (MyBP). In that period, the earth was about 4°C warmer than today. The world as we know it, with ice on both poles and glaciers on the highest mountains, came into existence as recently as about 2.5 MyBP.

By comparison with the previous history of the earth, the last 2 million years (the *Pleistocene*) were remarkably steady in climate, and significantly cooler on average. Data for the last 850,000 years (Fig. 1.1) indicate that over this time period, the average temperature of the Northern Hemisphere was about 13 + 2.5°C (Mitchell 1977). It is this cooler climate that has spawned most of present-day biotic life.

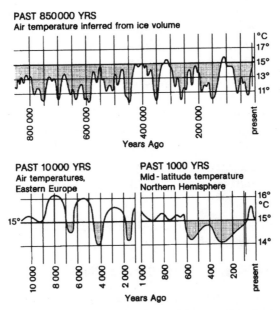

Figure 1.1   *History of mean air temperature as inferred from a variety of paleoclimatic indictors.*

The appearance of the first human beings on earth also coincides with the beginning of this period.

Within the Pleistocene, about 20 glacial and interglacial periods occurred that alternated roughly every 100,000 years. The warmest peak occurred about 125,000 years ago (the *Eem-Sangamon interglacial*), when it was about 2–2.5°C warmer than today.[10] Geological data suggest that during that time, the Antarctic ice shield had partially dissolved and sea levels were 5–7 m higher than today. Human beings existed solely in hunter–gatherer societies.

Within the primary, 100,000-year cycles of the Pleistocene, smaller oscillations occurred about every 20,000 years. The maximum of the last glacial period was about 20,000 yBP (years before present).

Based on these long-term patterns, the earth would be expected to gradually enter a new ice age over the next 10,000 years. However, this long-term trend expectation of an ice age has no bearing on the present concern over human-induced climate warming. That concern is over unprecedented climate warming within the next few decades or a century. Over that time horizon, which typically presents the outer bound of present policy discussions, the expected natural cycles would keep the earth solidly in an interglacial state, that is, near the 15°C mark at which we find ourselves today (Fig. 1.1).

The last of these shorter interglacials—a warm period the earth is still experiencing—began about 12,000 yBP. This period (the *Holocene*) reached its warmest phase about 6000 yBP, when the earth was about 1.5°C warmer than today. It is in this period that the first agricultural civilizations appeared.

Still more recently, a medieval warm period occurred around 1000 A.D., which was about 1°C warmer than today. The latest warm period was between 1800 and 1940, when temperatures rose by about half a degree (Hansen and Lebedeff 1988).

2. Anticipated Global Climate Changes

Anticipated Warming. The degree of climate change the world is likely to experience depends on future atmospheric greenhouse gas concentrations. Warming predictions also depend on the climate sensitivity assumed in the mathematical models used for calculating

---

[10]See Flohn (1983) and Barnola et al. (1987). In this and the following temperature ranges, the lower number refers to warming relative to the 1951–1980 mean, while the higher figure refers to conditions around 1860.

warming. According to established scientific consensus, a doubling of carbon dioxide levels or equivalent changes in atmospheric composition would raise the mean surface temperature on the globe anywhere from 1.5–4.5°C (CDAC 1983). The currently most widely used models predict a narrower range of 3–5.5°C (Dickinson 1986). When a broader range of possible feedback effects (see Section C) is added, average warming from doubled CO_2 could be as high as 6.3–8°C or more (Lashof 1989).[11]

This global average surface warming would be unevenly distributed. For instance, if low latitudes were to experience a warming of about 2°C, high latitudes would likely see as much as 4–10°C increase during the winter months.

Unprecedentedness, or Climatic Throwback. The enormity of these changes can only be grasped if we compare them to the climate history of the earth:

- A 1–1.5°C global average warming would represent a climate not experienced since the Holocene period at the beginning of agricultural civilization (some 6000 yBP).
- A 2–2.5°C warming would represent a climate not experienced since the Eem-Sangamon interglacial period some 125,000 years ago. At that time, human communities existed as hunter–gatherer societies, and the West Antarctic ice shield seemed to have partially disintegrated, raising sea levels by up to 5–7 m.
- A 3–4°C warming would represent a climate not experienced since human beings appeared on earth some 2 million years ago. The last time the earth was this warm was 5–3 million years ago, in the Pliocene period.
- A global average warming of 5°C or more would mean a climate not experienced for tens of millions of years, when there were no glaciers in the Antarctic, Iceland, and Greenland or on mountain ranges like the Sierra Nevada.

Figure 1.2 summarizes this comparison in the form of the degree of unprecedentedness, or *climatic throwback*: the number of years each

---

[11]However, the same study suggests that the different feedbacks might well produce stabilizing interactive dynamics rather than just being linearly additive. They also would probably lose much of their potency as substantial warming is realized and the absorption bands of greenhouse gases saturate.

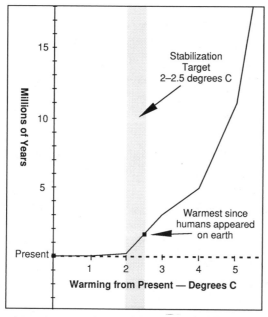

Figure 1.2 *Climatic throwback from global warming.*

additional degree of warming would take us back in human history and geological time to find comparable conditions. This throwback can be seen as a crude proxy for the risks from global warming.

Two qualifications are in order. First, the availability of data, and therefore our knowledge, about the environmental conditions during previous warm periods drops off steeply as we go further back in geological time. Second, some of the evidence in the earth's historic climate record suggests that the climate system, once sufficiently disturbed, may find a new equilibrium quickly and abruptly, without the gradual response in proportion to rising concentrations that current models predict (Broecker 1987).

Secondary Climate Changes. One major effect global warming will have is to create warmer oceans. The warmer oceans will evaporate more moisture. The excess moisture in the atmosphere will make the climate more humid and wetter overall. And global circulation patterns in the oceans and atmosphere will be affected.

While the uncertainties surrounding more specific predictions of impacts are often very large, the following geophysical and biospheric changes are expected:

- Rising sea levels by at least 0.5 – 1.5 m over the next few decades, and by as much as several meters over the longer term.
- Lower snowpacks and receding glaciers.
- Shifts in ocean currents and changed precipitation patterns in all regions.
- More frequent occurrences of weather conditions now considered extreme.
- Increased storms, floods, and avalanches, and significant seasonal changes in the availability of water runoff.
- Loss of soil moisture due to increased evaporation, and an increase in the duration and frequency of heat waves and drought conditions.
- Reduction in precipitation in the midlatitude continental regions of North America and Eurasia.
- More stagnant air masses, and displacement of high pressure systems.
- Changes and reorganizations of natural and agricultural ecosystems.

3. Impacts on Society

A growing amount of research is being done on how climate changes might affect human society, but the prediction of many impacts remains highly speculative at this time. Below, we summarize some of the most likely practical consequences of climate warming, along with risks of impacts that are less well established. A more detailed review can be found, for example, in Kates et al. (1985), Parry et al. 1988, and Smith and Tirpak (1988), the last of whom specifically discuss the United States.

The likelihood and extent of many impacts, particularly those having to do with the feedback effects of climate warming on geophysical and ecological processes, are highly uncertain at this time. Often, there are countervailing feedback forces at work. These can make a determination of the net direction of impacts difficult (Lashof 1989). However, prudent policy-making should be particularly concerned with the downside risks, and not rely on the most optimistic interpretations of available evidence.

Food Security. Human history is replete with famines induced by deterioration in regional climates, and even highly mechanized and chemicals-intensive agriculture is critically dependent on climate.[12] Under favorable conditions, increased CO_2 levels and higher temperatures could increase agricultural yields. But many crop yields are delicately dependent on a particular mix of temperatures, soil conditions, and rainfall patterns. High latitude regions that could in principle become available for agriculture may not provide such favorable conditions. Furthermore, a number of weed plants seem to be more efficient in utilizing atmospheric carbon dioxide than are crops. Warmer weather would also encourage the spread of certain plant pests (Bach et al. 1981).[13]

A recent study by the U.S. Environmental Protection Agency predicts that warming of several degrees would force major redistributions of cropping zones and changes in farming practices (Smith and Tirpak 1988). The study found that while some areas might gain a few tens of percent in yields, other areas could suffer reductions of 50 percent. Drier conditions are expected for the grain belts in the U.S. Midwest. Semiarid regions would be particularly affected. The same goes for the wheat-growing Kazakstan region of the Soviet Union.

Impacts on agricultural productivity could be especially severe in developing countries. Notably in the semiarid regions, small irregularities in climate are sufficient to create major disruption. Heat stress could also severely reduce rice plant reproduction in the world's rice-growing regions. Such impacts could have disastrous effects on poor rural subsistence farmers in the developing world. Unlike farmers in industrialized nations, most Third World peasants do not have modern fuels and machinery to help them deal with fundamental changes in their environment. A major productive asset of these farmers is their intuitive knowledge of the local climate, and the cultural "software" that has been built around this knowledge. Global warming could make both worthless.

Reduced yields and less-than-needed yield improvements in the developing countries, combined with higher food prices worldwide and possibly the loss of surplus production and reserves in industrialized countries, could lead to more suffering in the Third World. In the

---

[12]The 1988 drought in the United States was a reminder of this dependence. For the first time in decades, the largest grain exporter in the world failed to produce more wheat than that needed to meet domestic demand.

[13]Again, the 1988 U.S. drought is a case in point. The unusually hot conditions resulted in the infection of significant portions of the corn harvest with aflatoxin-producing molds. Aflatoxins are potent carcinogens.

international markets, scarce supplies would go to those who could pay the highest price. Such shifts in fortunes could bring geopolitical instability and international polarization similar to that produced by the oil crises of the 1970s.

Impacts on Forests and Species Diversity. While most attention has been given to impacts on agricultural systems, important disruptions are expected for relatively unmanaged ecosystems. Where soil conditions and rainfall permit, global warming would allow forests to extend upward in altitude and poleward in latitude. Unfortunately, the rate of warming the world might experience over the next number of decades could far outstrip the capacity of forests to migrate. A several degree warming over a 100-year period would greatly exceed natural rates of change, pushing forests poleward by 2.5 km per year, compared to less than 1 km per year in fast-migration tree species (CEC 1986).

If forests cannot keep up with this rate, the result would be rapid dieback of existing forests while new species would take root much more slowly. The rapid appearance of acid rain damage in the forests of central Europe illustrates the vulnerability of these ecosystems, and continued air pollution stresses from fossil fuel consumption could further aggravate these impacts. The result could be a period of many decades in which forest cover and/or forest productivity in the mid-latitudes would be greatly diminished.

Climate change might not just disrupt midlatitude forests, but could also lead to serious damage to tropical forests, due to greater evaporation and changes in the regional distribution of rainfall. This, in turn, could exacerbate current losses in species diversity. Most of the world's biological species are found in these forests, and human-induced deforestation is already producing species losses at an alarming rate.

Land Use and Human Settlements. Warming of several degrees could within the next 50 to 100 years result in a sea-level rise of 0.5 to 1.5 m. Global warming could also eventually break up the West Antarctic ice shield and lead to a sea-level rise of several meters, though this process would probably take several hundred years.

Even a modest sea-level rise would threaten the coastal settlements in which half of humanity lives. In the United States alone, an estimated 12 million people—close to 5 percent of the population—could become homeless. High tides and storm surges would penetrate further inland. Salinity would move upstream, penetrating into groundwater and bays, and forming inland salt lakes in many areas. Rich farmland in coastal river deltas would be lost unless expensive dikes were built. Developing countries, particularly in Asia, may find the cost of such measures prohibitive.

Economic impacts would be aggravated by the fact that many major infrastructures and industrial investments are located near the shore. Airports, waste dumps, harbor facilities, locks, bridges, drainage systems, irrigation systems, water treatment plants, chemical factories, and power plants all would require protective investments or rebuilding to protect them from flooding. Smith and Tirpak (1988) estimate that protection against a 1-m rise in sea levels would cost a cumulative $123–175 billion (1985 dollars) by the year 2100 for the United States alone. Again, Third World countries would be especially hard hit.

Freshwater Supplies. Warming could reduce stream flows and increase pressure on groundwater supplies in many regions, while worsening the pollution from waste discharge into smaller flows. For example, it is estimated that viable farm acreage in the arid regions of the United States could be reduced by one third (Waggoner 1989). Regions and nations sharing watersheds could experience conflicts in the form of water wars.

Other Impacts. Other impacts could involve human health risks, due to extreme heat stress and the more vigorous transmission of tropical and other diseases over larger areas; increases in energy consumption for air conditioning (Linder and Inglis 1988);[14] losses in hydropower availability; and losses in revenue from tourism and fisheries.

Planning Uncertainty. A major impact, and one that is frequently overlooked, is the great increase in planning uncertainty. Climate change could be producing impacts rapidly in some regions but slowly in others, and at different speeds during different phases of the warming process. Such uncertainty would impact all levels of human activity, including the very investments aimed at adaptation to climate change—be it the planning of flood control systems, adjusting hydro dams and irrigation systems to irregular runoff, or selecting food crops for changing growing seasons. Society might find itself engaged in a

---

[14]Linder and Inglis estimate that annual electricity consumption in the United States would increase −0.5 to 2.7 percent per degree Celsius, depending on the local climate and on the saturation of electric air conditioning and heating equipment. Their analysis neglects, however, that modern high efficiency, passive cooling, and cool storage technologies could minimize or eliminate the need for additional active heating or air conditioning.

constant treadmill, trying to catch up with perpetual change in an environment that no longer works.

E. ANALYTIC FRAMEWORK FOR THE PREVENTION PARADIGM

1. Conventional Analyses: Warming Fates

Past analyses of the climate threat have for the most part focused on translating widely accepted trend-based projections in greenhouse gas emissions into estimates of global warming. In this manner, policymakers were presented with menus of "warming fates" that left little for them to do except fund more research. Climatologists, in turn, saw their main task as conveying to policymakers the likely consequences and uncertainties of the various warming predictions.

As a basis for their calculations, analysts would use observed atmospheric concentration trends or government and industry projections for energy use and CFC production. Without exception, these projections indicate inexorably growing emissions of greenhouse gases. In effect, the question of what climate warming should be considered tolerable or unavoidable has remained unanswered.

With growing public support for a policy paradigm of prevention, several governments have commissioned analyses of options to slow and, where possible, eliminate the anthropogenic driving forces of global warming. This reorientation toward prevention necessitates an entirely different analytical exercise, at the center of which are detailed technical and policy options to reduce greenhouse gas emissions.[15]

2. Prevention-Oriented Climate Analysis:
The Warming Limit Approach

The analysis in this book differs from past analyses in terms of both objectives and methodology:

- The *objective* of our analysis is to define a course of preventative action that would limit the impacts and risks from global warming

---

[15]An early implementation of such a prevention-oriented approach was the study by Lovins et al. (1981). Two major policy studies, which were prepared independently and more or less in parallel with this study, are those of the U.S. Environmental Protection Agency (Lashof and Tirpak 1989) and by UNEP and the Beijer Institute (Swart et al. 1989).

to an unavoidable minimum. We call the class of emission paths and policy scenarios that achieve this goal *toleration scenarios.*

- Our *method* of analysis is to iteratively translate explicit climate stabilization goals into policy targets in the fields of energy, agriculture, forestry, foreign aid, pollution control, chemicals regulation, and so on.

In our approach, three concepts are of central importance:

- A global average surface *warming limit.*
- Global *cumulative emission budgets* that are compatible with the warming limit.
- Global *emission reduction milestones* that are compatible with the emission budgets and warming limit.

We have dubbed this approach the *warming limit/emission budget/ reduction milestone method* (WERM).

Why an Explicit Warming Limit Is Important. All prevention-oriented approaches must ultimately rely on policy targets that take, directly or indirectly, the form of a ceiling on global surface temperature warming. A global warming limit has been the subject of recent international scientific discussions (Jaeger et al. 1988), and has been further explored by the FRG Enquête-Kommission (1988) and Swart et al. (1989). Our study is the first to implement such a warming-limit approach, going the full length from target setting to a detailed specification of required emission curtailments. Our goal is both to demonstrate the analytic viability of such an approach, and to establish initial orders of magnitude. These will hopefully be refined as tools and data improve.

To date, most analyses of preventative action have circumvented the issue of an explicit warming limit, turning directly to practical targets for the curtailment of greenhouse gas emissions. Such practical targets, in turn, are developed on the basis of economic, social, and technical feasibility.[16] This pragmatic approach is eminently sensible, but it has an important drawback: lacking a common climate target for risk abatement, there is the danger of inadvertently slipping into definitions of feasibility that are more a reflection of the current status quo than of climate risks.

A warming limit does not have to be in contradiction with a prag-

---

[16]A comprehensive analysis of this kind is found in Lashof and Tirpak (1989).

matic approach. On the contrary, if used iteratively, it could boost the transparency and quality of the international dialogue and ultimately lead to a more satisfactory pragmatism. We illustrate this iterative approach in our study: starting with a warming limit derived solely from considerations of climate risk, we evaluate its consequences on society in an explicit, transparent manner, assessing its logistic, economic, and other implications. Exploring climatic requirements and response options in detail, and without excessive regard for the present state of affairs, provides an incentive to investigate possible responses fully, and to scrutinize assumptions about feasibility. A practical compromise between the desirable and the feasible can then be made through informed public discourse.

Making Climate Targets Useful for Policy. Any target formulated on the basis of climatic parameters such as surface warming needs to be translated into equivalent targets that can be practically used by decision makers. This operationalization should satisfy three criteria:

- The warming limit should be accompanied by a comprehensive assessment of the potential for reducing greenhouse gas emissions. This assessment should indicate how much these emissions will need to be reduced to stay within the warming limit.
- A method must be provided for translating the global warming limit into an equitable allocation of remaining permissible greenhouse gas emissions among the world's nations and regions.
- The above translation should rely on simple computational procedures that do not invite excessive controversy over computational and measurement issues.

These three aspects, which are interrelated, guide our methodological discussion.

Conceptual and Computational Challenges of a Warming Limit Approach. A target ceiling on global surface temperature rise brings with it important computational and conceptual issues that are inherent to the physics of the climate system and the greenhouse effect. These fall into two categories: inherent climate system complexities, and uncertainties in available data and knowledge. The former require sufficiently sophisticated modeling tools, the latter more research. Where resolution through research is infeasible, and as an interim approach while research is being carried out, normative agreements on how to treat such uncertainties will have to be negotiated.

Some of the inherent complexities were discussed in Section C. They include

- One and the same absolute amount of cumulative emissions will cause different surface warming impacts depending on how these emissions are distributed over time. Conversely, one and the same warming increment for a particular trace gas could be compatible with varying amounts of emissions, depending on their timing.
- The atmospheric concentration that results from emissions of a greenhouse gas and that governs its radiative forcing may likewise depend on the distribution of such emissions over time.
- In the case of radiatively and chemically active greenhouse gases, the concentrations are also dependent on other emissions with which they interact chemically.
- The radiative forcing from a given atmospheric concentration of a greenhouse gas depends somewhat on the concentration of other greenhouse gases, due to overlapping absorption bands.
- A warming limit can only be translated into a total radiative forcing limit and corresponding *sets of* emission reduction targets. Within these sets, reductions for one particular trace gas could be varied on the basis of trade-offs. To make one-gas targets meaningful, it is necessary to state explicitly and clearly what emission assumptions are being assumed for all the other trace gases.

This means that the warming impact of an increment of greenhouse gas emissions can be accurately assessed only in the context of combined emission scenarios covering all greenhouse gases over extended periods of time.

The second category of challenges includes the following:

- Because of the uncertainty in climate sensitivity, permissible greenhouse gas emissions under one and the same warming limit will vary depending on the climate sensitivity used to model the climate system (model climate sensitivity).
- Because of uncertainties about the rate and mechanisms of CO_2 removal from the atmosphere, one and the same emission path can lead to different concentrations, depending on the carbon-cycle model used.
- For some other important trace gases, notably CH_4 and N_2O, the sources and sinks are so little understood that it is difficult to correlate emission reductions with future concentrations reliably.

Treatment of Uncertainty in Climate Sensitivity. Of the three types of uncertainty mentioned, climate sensitivity is the most significant. In fact, it is so important that any warming limit must ultimately be specified in terms of a reference climate sensitivity to be operational.

In this book, the range of climate sensitivities is made an explicit element in the iterative determination of a warming limit. Our philosophical approach is that of risk minimization as practiced in other fields of public protection, such as dealing with toxicity, ozone depletion, and carcinogenic substances. This means that we choose emission limits that would be compatible with the high end of the range of currently accepted climate sensitivities. The required emission curtailments are then compared with what is practically feasible. We devote Chapter 6 to defining the meaning of practical feasibility in greater detail (see also below and Chapter 2).

3. The Most Rigorous Approach: Allocating the Warming Pie

Once a maximum permissible warming increment has been determined and a reference climate sensitivity specified, the ideal approach would be to deal with the allocation problem directly at the temperature level itself: shares of the global increment could be allocated among the world's nations on the basis of need and equity. (Formulas for such allocation are discussed in Chapter 5.) Permissible emissions for each country would then be derived on the basis of comprehensive climate modeling.

The method of implementation could be as follows:

- Nations and regions prepare long-term emission plans covering all greenhouse gases over a period of 50 to 100 years. These plans would specify actual investments for the near-to-medium term, and simple pledges without specific implementation details for reductions in the long term. A specific emission path would thus be proposed for each greenhouse gas, in the form of a series of emission rate milestones.
- A warming impact assessment of these plans would be done on the basis of some standardized modeling framework,[17] perhaps by a specially created international agency. Filings with this agency

[17]Examples are the Atmospheric Stabilization Framework (ASF) of the U.S. Environmental Protection Agency (Lashof and Tirpak 1989) and the IMAGE framework of the Dutch National Institute for Public Health and Environmental Protection (RIVM).

would be somewhat akin to getting a smog check for one's car, only that in this case it would be nations getting a "greenhouse check."

- The plans would be evaluated in terms of their radiative forcing or warming impact. Evaluations could be done by treating each plan as an incremental change in a global reference scenario, or using approximate concepts, such as a temperature increase potential for each type of trace gas emission (Swart et al. 1989). If in excess of the allowed temperature increment, the emission reduction plan would be corrected until in agreement with the allocated warming increment.
- To provide monitoring and flexibility, updated and revised plans would be filed on a periodic basis. At each filing, remaining warming increments would be recalculated, and corresponding revisions would be made in the greenhouse gas reduction milestones.

The advantage of this approach is that it calculates emission curtailment targets with the maximum scientific exactitude current knowledge allows, notably by determining warming impacts on the basis of the specific sequence and timing of emissions. Given the memory of the climate system, this is a potentially important feature. The approach also would allow nations complete freedom to devise greenhouse gas (GHG) curtailment programs matched in timing and mix of measures to their own specific circumstances.

Eventually, this approach may become practically feasible. However, at present this scientifically optimal approach is less ideal in other respects. At the present stage, such an approach could prove premature and overly elaborate. Most disturbing is the discrepancy between the sophistication of the modeling tools that would be brought to bear, and the crudeness of data and large uncertainties in the sensitivity of the climate and in some GHG cycles. The nation-by-nation performance and periodic repetition of elaborate modeling-based greenhouse checks could be problematic. It could invite incessant procedural challenges and haggling over details, in effect undermining the international agreements that brought about the process in the first place. Periodic revisions in scientific knowledge could exacerbate such pressures.

4. The Opposite Extreme: Single-Point (Near-Term) Reduction Targets

Recent calls for climate-stabilizing action have taken the form of percentage reduction targets for carbon releases. For example, the

1988 Toronto conference called for a 20 percent emission reduction by 2005. Such target proposals are a welcome change from more complacent attitudes. Simple, short-term goals are particularly conducive for engaging decision makers. Also, by focusing on carbon dioxide only, they avoid the issue of trade-offs among GHG reduction targets.

At the same time, percentage reduction targets for one greenhouse gas and a single (near-term) point in time are scientifically unsatisfactory and plainly insufficient for characterizing a climate-stabilization strategy. Energy investments, for example, shape future patterns of (fossil) energy consumption for up to 50 years, not just over 10 or 15 years. Specific reduction milestones should be developed in the context of at least an approximate long-term global emission path, using climate modeling. The shape of emission trajectories in later years should be considered in setting targets for earlier years. In general, the present short-term proposals are not sufficiently grounded in the physical dynamics of the greenhouse problem. Equally important, simple, undifferentiated percentage reduction targets suggest a "one size fits all" approach that could prove unnecessarily rigid and inequitable (see also Chapters 5 and 6).

5. A Viable Approximation: Cumulative Emission Budgets

The intermediate approach used in this book centers around the concept of cumulative emission budgets as proxies for the global warming limit. This proxy allows shifting the international allocation issue to the level of emissions, by assigning each country and world region a certain share of the global trace-gas budget.

The procedure described here focuses on carbon dioxide, for the following reasons: carbon dioxide is the dominant contributor to global warming (see Chapter 2); its anthropogenic sources are much better understood than those of N_2O and CH_4; and unlike for the chlorofluorocarbons, there is no international control process under way at this time. These factors explain the overall energy policy orientation of this study.

The steps in our analysis are as follows:

1. Establish a reference period over which the issue will be analyzed. We choose the period until 2100, which is sufficiently long to allow the study of delayed effects on the climate system.
2. Set an initial warming limit as a working hypothesis. This limit can be iteratively corrected at any point along the course of the overall analysis.

3. Compare the chosen warming limit with the estimated unrealized warming (warming commitment) from past human activity, which is currently concealed due to the thermal inertia of the oceans. Use the results to eliminate from the range of uncertainty in climate sensitivities those sensitivities, if any, for which the warming limit can no longer be realized.

4. Explore what sets of concentration ceilings and emission scenarios for the major greenhouse gases would be compatible with the target warming limit. Use published scenarios and both equilibrium and transient climate modeling[18] to establish which previously proposed technological policies and economic and population developments, if any, are potentially compatible with the warming limit. If none are, calculate what emission paths and radiative forcings would be.

5. Extract from these modeling results an approximate trend curve and band of radiative forcing or equivalent CO_2 concentrations that can be used as a representation or proxy for the warming limit. Develop this radiative forcing limit or concentration band as a function of a range of model climate sensitivities. Set a total radiative forcing limit using the highest climate sensitivity for which climatic targets would still be feasible (goal: risk minimization).

6. Assess the potential of various policy measures and technological fixes for controlling greenhouse gases other than fossil carbon dioxide emissions, taking into consideration current developments in CFC control and future economic and population growth in the developing countries. Evaluate whether the emission reductions identified or required in steps 4 and 5 could be achieved on the basis of identifiable measures. Make some heroic (but explicit) assumptions about the less understood trace-gas sources.

7. Subtract from the total radiative forcing (equivalent CO_2 concentration) obtained in step 5 the unavoidable equivalent CO_2 concentrations of nonfossil trace gases as defined in step 7. This procedure defines an approximate maximum concentration limit for fossil carbon dioxide under the warming limit.

8. Calculate bands of cumulative fossil carbon emissions in the period till 2100 that would be permissible under this concentration ceiling, using various carbon-cycle models.

---

[18]See Chapter 2 for definitions.

9. Narrow the range to the minimum global fossil carbon budget that would appear practically feasible. At this stage, the limit of practical feasibility is a working hypothesis to be verified through the subsequent analysis.

10. Allocate the global fossil carbon budget among industrialized and developing countries, using an iterative assessment of international equity goals and practical feasibility.

11. Develop fossil phaseout scenarios and reduction milestones for industrialized and developing countries based on the constraint of fossil carbon budgets. Iterate with logistic constraints, such as those associated with starting and maintaining large-scale programs to restructure energy-producing and energy-consuming capital stocks. If necessary, iterate with the carbon budget analysis in step 9.

12. Perform least-cost analysis of options to meet the required reductions, taking into account the potentials of the various low-fossil or nonfossil resources and future economic and population growth. Evaluate the results in terms of the net cost and economic feasibility of climate stabilization as outlined in Section A above.

Evaluation. The concept of a cumulative fossil carbon budget has a number of attractions: it allows a simple, transparent treatment of the allocational issue in terms of emission rights; it provides individual nations and regions the same kind of flexibility in devising their reduction plans as the warming increment approach; it is relatively easy to monitor on the basis of existing data-gathering mechanisms; and in covering the long term, it avoids the frequent adjustments required in the warming increment approach.

An inherent simplification in this approach is that cumulative emission budgets are blind to the effect of the timing of releases on the warming obtained. Alternative paths for the reduction of fossil carbon emissions, though cumulatively equivalent, may still produce different warming impacts in the climate system.

This may be a reasonable price to pay for elegance and simplicity, but even from a climate-modeling point of view, such budgets can be an acceptable approximation. If the variability in the timing of global emission reductions over the period till 2100 is not excessive, warming differences resulting from the inertial dynamics of the climate system will not be large. As will be shown in Chapter 6, such a limited variability can be expected: because there are significant logistic constraints in reorganizing the world's energy systems, a risk-minimizing (i.e., tight) fossil carbon budget results in significant constraints on the timing of the global fossil phaseout.

6. Target Limits for the Rate and Magnitude of Climate Warming

What warming limit should guide climate-stabilization policy? Ultimately, this question requires a normative answer. However, several kinds of science-based indicators should be used as inputs for this determination. Our discussion in Section D suggests that both *rates* of change and *absolute limits* should be incorporated in a climate-stabilization target.

Rate of Climate Change. The degree of risk associated with climate change is directly proportional to the rate of warming. The faster the change, the more likely it will outstrip the adaptive capacities of natural ecosystems, and the greater the risk of impacts such as massive forest dieback. Of particular interest are the rates of temperature change during the last several thousand years in which humanity has relied on agriculture for food production.

In the period from the waning of the last ice age some 10,000 years ago to the establishment of the present interglacial epoch, the average rate of temperature change was no more than $0.01-0.02°C$ per decade, though much higher during small fluctuations.[19]

Data on the ability of trees to migrate suggest a limit of $0.1°C$ per decade (Jaeger et al. 1988). This rate is about ten times faster than the natural average rate of temperature change seen from the end of the last ice age to the present, but is significantly less than what current trends in greenhouse gas emissions are calculated to produce (see Chapter 2).

If we were to limit the rate of warming on the basis of tree migration and historic precedent, the global warming limit for 2100 would be about $1.1°C$ ($1.6°C$ relative to 1860). As extreme upper limit numbers, a range of $1.5-2°C$ could be used.

Absolute Limit. An absolute limit on warming could be inferred from past levels of sea-level rise. Sea-level rise in the neighborhood of 1 m or so might still be manageable at enormous cost, but a $5-7$ m rise as during the Eem-Sangamon interglacial would surely be devastating.

---

[19]During the Little Ice Age and the recent warming between 1880 and 1940, temperatures appear to have changed at an average rate of about $0.05°C$ per decade or more.

On that basis, a plausible warming limit for the next several hundred years would be $2-2.5°C$.[20]

We propose the degree of climatic throwback as an indicator of climate risk. Figure 1.2 showed that temperature changes much above $2-2.5°C$ drastically increase the amount of throwback, and consequently magnify the risks from global warming. A $2-2.5°C$ ceiling also has an intuitive appeal: most of the earth's current gene pool, including *Homo sapiens*, evolved under the temperature conditions prevailing over the last 2 million years. Never since the appearance of the first human being has the world been any warmer than this ceiling. Any additional warming would lock humanity and most other species out of the climatic conditions under which they were raised. While we are not prisoners of our biological past, our growing understanding of human evolution and its dependence on functioning natural ecosystems suggests that we are far from free of it. We should take note of this transgression into uncharted territory.

Summary. Our discussion shows that the most tangible guideline would be to limit warming on the basis of the adaptive capacity of forests. This guideline, which was proposed at the Villach and Bellagio meetings of the UNEP/WMO Climate Impacts Program (Jaeger et al. 1988), would yield a limit of $1.1-1.6°C$ for the period until 2100. Targets based on indices of absolute warming are less easily derived for this period, but a limit of $2-2.5°C$ appears plausible and defensible. The German Meteorological and Physics societies call for a warming limit of $1-2°C$ (DMB/DPG 1987), and the same limit has been suggested by the FRG Enquête-Kommission (1988).

For the purpose of this study, we define climate stabilization as follows:

- The average rate of global warming is to be limited as closely as possible to $0.1°C$ per decade.
- The currently rising trend of global average surface temperatures from human activity is to be reduced to very small rates (substantially below the mark of $0.1°C$ per decade) by the end of the next century.
- As an outer limit, temperature rise is to be limited to $2°C$ relative to the present ($2.5°C$ relative to 1860).

[20]However, because such a sea-level rise would involve the partial dissolution of the Antarctic ice shield, it would most likely take several hundred years. Thus, a limit on sea-level rise would allow excursions into much higher temperature ranges in the interim.

It should be reiterated that these are normative targets. But even readers who might disagree with our risk assessment should benefit from the subsequent analysis. Implementation of the analytical approach demands and organizes information in a manner that will facilitate quantitative variations based on other targets.

7. Emission Budgets and Reduction Schedules

In this section, we illustrate schematically how emission budgets and reduction schedules are related to each other in the WERM framework. In Chapter 6, we use this analysis for evaluating, among other things, how the Toronto target fits into a comprehensive view of climate stabilization.

The basic relationships are shown in Figure 1.3. The trace-gas concentration limit is expressed in the form of an emission budget for the period during which climate stabilization is to be implemented (taken here as the period from 1985 to 2100). This emission budget is shown in the form of a rectangle, with the height equivalent to the base-year emission rate, and the length equivalent to the years it would take to use up the cumulative emission budget at that rate.

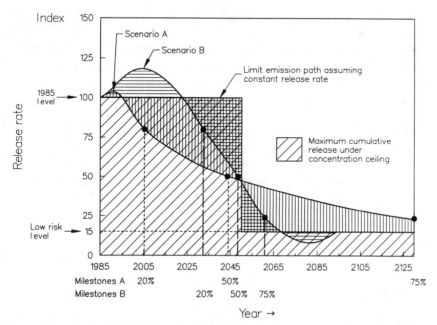

Figure 1.3 *Climate-stabilization targets: relationship of concentration ceilings, emission paths, and reduction milestones.*

Also shown is a low-risk emission level at which concentrations would not increase, due to the absorption capacity of the more inertial components of the chemical cycle. Unfortunately, the emission levels at which concentrations would no longer rise can only be specified approximately, due to our limited understanding of the relevant global chemical cycles. The same uncertainties limit the accuracy with which a cumulative emission budget can be specified (see Chapter 4).

Two illustrative emission reduction scenarios are shown. While their shapes are different, both result in the same cumulative emissions.[21] In scenario A, the initial phaseout of emissions proceeds promptly, leading to a 20 percent reduction within about 20 years. This allows for a moderate speed of phaseout in later years. Thus, a 50 percent reduction is reached in about 60 years, and a 75 percent reduction in about 145 years.

Scenario B shows a delayed phaseout scenario. Here, it takes about 50 years to reverse current growth trends and achieve any reductions. By the time the early scenario had decreased emissions by 20 percent, this scenario has not even returned to base-year levels. As a result, the speed of reduction (the slope) now has to be much steeper to remain within the overall emission budget. The 50 percent reduction milestone now has to be reached 15 years later, and the 75 percent reduction target is only about 30 years away.

The graph illustrates that countries could reduce greenhouse gas emissions along flexible schedules so long as cumulative trace-gas emission budgets are observed. The degree of this latitude is, of course, limited by logistic and other constraints in phasing out trace-gas emissions.[22] These differ for each trace gas.

REFERENCES

Bach, W. (1982–1984), *Gefahr für unser Klima. Wege aus der CO$_2$-Bedrohung durch sinnvollen Energieeinsatz*, C. F. Muller, Karlsruhe, Germany. English version 1984: *Our Threatened Climate: Ways of Averting the CO$_2$*

---

[21]However, the concentrations reached could be somewhat different for each path if there is significant nonlinearity in the response of the chemical cycle to varying levels of emission rates. See Chapter 4 for an illustration of this phenomenon in the case of the carbon cycle.

[22]As mentioned, the fact that variations among emission trajectories are logistically limited is one reason for using the emission budget/milestone schedule approach as an approximation.

Problem Through Rational Energy Use, Reidel, Dordrecht, the Netherlands.

Bach, W., and H. J. Jung (1987). "Untersuchung der Beeinflussung des Klimas durch Spurengase mit Hilfe von Modellrechnungen," *Münstersche Geographische Arbeiten,* Vol. 26, pp. 45–64.

Bach, W., J. Pankrath, and W. W. Kellogg, eds. (1979), *Man's Impact on Climate,* Elsevier, Amsterdam.

Bach, W., J. Pankrath, and J. Williams, eds. (1980), *Interactions of Energy and Climate,* Reidel, Dordrecht, the Netherlands.

Bach, W., J. Pankrath, and S. H. Schneider, eds. (1981), *Food-Climate Interactions,* Reidel, Dordrecht, the Netherlands.

Barnola, J. M., D. Raynaud, Y. S. Korotkevich, and C. Lorius (1987), *Nature,* Vol. 329, p. 408.

Broecker, W. S. (1987), "Testimony before the U.S. Senate Subcommittee on Environmental Protection," Jan. 28, Washington, D.C.; see also *Nature,* Vol. 328 (1987), p. 123.

Brown, L. R., et al. (1988), *State of the World Report,* Worldwatch Institute, Washington, D.C.

Bruntland Commission (1988), *Our Common Future,* Report to the United Nations Commission on Environment and Development, United Nations, New York.

Budyko, M. I., A. B. Ronov, and A. L. Yanhin (1987), *History of the Earth's Atmosphere,* Springer-Verlag, Berlin.

CDAC (Carbon Dioxide Assessment Committee) (1983), *Changing Climate,* Natural Resources Council, National Academy Press, Washington, D.C.

CEC (1986), "Climate Change and Associated Impacts," *Proceedings of the Symposium on CO_2 and Other Greenhouse Gases: Climatic and Associated Impacts,* Brussels, Nov. 3–5, 1986.

Chamberlain, J. W., H. M. Foley, G. J. MacDonald, and M. A. Ruderman (1982), "Climate Effects of Minor Atmospheric Constituents," in W. Clark, ed, *Carbon Dioxide Review,* Oxford, 1982, pp. 253–277.

CHCN (1983), *Report on CO_2 Problem,* Committee of the Health Council of the Netherlands, The Hague, the Netherlands.

Dickinson, R. E. (1986), in B. Bolin, R. Doos, J. Jaeger, and R. A. Warrick, eds., *The Greenhouse Effect: Climatic Change and Ecosystems,* Wiley, New York, pp. 207–270.

Dickinson, R. E., and R. J. Cicerone (1986), "Future Global Warming from Trace Gases," *Nature,* Vol. 319, pp. 109–114.

DMG/DPG (Deutsche Meteorologische Gesellschaft/Deutsche Physikalische Gesellschaft) (1987), *Warnung vor drohenden weltweiten Klimaanderungen durch den Menschen,* Bad Honnef, FRG, 16 pp.

Enquête-Kommission (1988), *Erster Zwischenbericht, Enquête-Kommission*

Vorsorge zum Schutz der Erdatmosphaere of the FRG Parliament, Deutscher Bundestag, Bonn, Nov. 2.

Flohn, H. (1983), "Das CO_2-Klima Problem," *Geographische Rundschau*, Vol. 35, No. 5, pp. 238–247.

Goldemberg, J., T. Johannson, A. K. N. Reddy, and R. H. Williams (1988), *Energy for Sustainable Development*, World Resources Institute, Washington, D.C.

Hansen, J. E. (1989), Presentation at the American Geophysical Union Fall 1989 Meeting, San Francisco, Dec.

Hansen, J. E., and S. Lebedeff (1988), "Global Surface Air Temperatures: Update Through 1987," *Geophysical Research Letters*, Vol. 15, pp. 323–326.

Hengeveld, H. G. (1987 and other years), *Understanding CO_2 and Climate*, Annual Report 1986, Canadian Climate Centre, Ottawa.

Jaeger, J., et al. (1988), *Developing Policies for Responding to Climate Change*, World Climate Program Impacts Studies, World Meteorological Organization and United Nations Environment Program, Washington, D.C., April.

Kates, R. W., J. H. Ausubel, and M. Barberian eds. (1985), *Climate Impact Assessment: Studies of the Interactions of Climate and Society*, Wiley, New York.

Krause, F., H. Bossel, and K. F. Müller-Reissmann (1980), *Energiewende*, S. Fischer Verlag, Frankfurt.

Lashof, D. A. (1989), "The Dynamic Greenhouse: Feedback Processes That May Influence Future Concentrations of Atmospheric Trace Gases and Climate Change," *Climatic Change* (forthcoming); Revised draft, Jan. 1989.

Lashof, D. A., and D. A. Tirpak, eds. (1989), *Policy Options for Stabilizing Global Climate*, Draft report to the U.S. Congress, U.S. Environmental Protection Agency, Office of Policy, Planning, and Evaluation, Feb.

Linder, K., and M. Inglis (1988), *The Potential Impacts of Climate Change on Electric Utilities: Regional and National Estimates*, Report of ICF to the U.S. Environmental Protection Agency.

Lovins, A. B., H. Lovins, F. Krause, W. Bach (1981), *Least-Cost Energy: Solving the CO_2-Problem*, Brick House, Andover, MA. (German version: *Wirtschaftlichster Energieeinsatz: Lösung des CO_2 Problems*, Karlsruhe, FRG, 1983.)

MacDonald, G. (1989), Presentation at the IEA Expert Seminar on Technologies for Controlling Greenhouse Gases, OECD, Paris, Apr. 12–14, MITRE Corporation, McLean, VA.

Mitchell, J. M., Jr. (1977), "Records of the Past, Lessons for the Future," in *Proceedings* of the Symposium on Living with Climate Change, Phase II, MITRE Corporation, McLean, VA, pp. 15–25.

Olivier, D., H. Miall, F. Nectoux, M. Opperman (1983), *Energy-Efficient Futures: Opening the Solar Option*, Earth Resources Research Ltd., London.

Parry, M. L., et al. (1988), *The Impact of Climatic Variations on Agriculture*, Vol. 1: *Assessments in Semi-arid Regions*, Vol. 2: *Assessments in Cool Temperate and Cold Regions*, Reidel, the Netherlands, Kluiver Academie.

Ramanathan, V., et al. (1985), "Trace Gas Trends and Their Potential Role in Climate Change," *Journal of Geophysical Research*, Vol. 90, pp. 5547–5566.

Ramanathan, V., et al. (1987), "Climate-Chemical Interactions and Effects of Changing Atmospheric Trace Gases," *Reviews of Geophysics*, Vol. 25, No. 7, pp. 1441–1482.

Raval, A., and V. Ramanathan (1989), "Observational Determination of the Greenhouse Effect," *Nature*, Vol. 342, pp. 758–67.

SCEP (1970), "Report of the Study of Critical Environment Problems," in *Man's Impact on the Global Environment*, MIT Press, Cambridge, MA.

Schlesinger, M. E. (1986), "Equilibrium and Transient Climatic Warming Induced by Increased Atmospheric CO_2," *Climate Dynamics*, Vol. 1, pp. 35–51.

Schneider, S. H. (1989), "The Greenhouse Effect: Science and Policy," *Science*, Vol. 243, No. 4892, (Feb. 10), pp. 771–781.

Schneider, S. H., and R. Londer (1984), *The Coevolution of Climate and Life*, Sierra Club Books, San Francisco.

Seidel, S., and D. Keyes (1983), "Can We Delay a Greenhouse Warming?" U.S. Environmental Protection Agency, Washington, D.C.

SMIC (1971), "Report on the Study of Man's Impact on Climate," in *Inadvertent Climate Modification*, MIT Press, Cambridge, MA.

Smith, J. B., and D. Tirpak, eds. (1988), *The Potential Effect of a Global Climate Change on the United States*, Draft report to Congress, Oct.

Sundquist, E. T., and W. S. Brocker, eds. (1985), *The Carbon Cycle and Atmospheric CO_2: Natural Variations Archean to Present*, American Geophysical Union, Washington, D.C.

Swart, R., H. de Boois, and P. Vellinga (1989), *The Full Range of Responses to Anticipated Climatic Change*, UNEP and the Beijer Institute, Apr.

Titus, J. G., ed. (1986), *Effects of Changes in Stratospheric Ozone and Global Climate*, Vols. I–IV, U.S. Environmental Protection Agency/UNEP, Washington, D.C.

UNEP (1987), Expert working paper submitted to the Enquête-Kommission, Bonn, FRG. United Nations Environment Program, New York/Nairobi.

U.S. DOE (1985), *State-of-the-Art-Reports*, DOE/ER-0235 to 0239, U.S. Department of Energy, Washington, D.C.

Waggoner, P. E., ed. (1989), *Climate Change and U.S. Water Resources*, Wiley, New York.

Wang, W. C., et al. (1986), "Trace Gases and Other Potential Perturbations to Global Climate," *Reviews of Geophysics*, Vol. 24, No. 1, pp. 110–140.

Wuebbles, D. J., and J. Edmonds (1988), *A Primer on Greenhouse Gases*, Report to be published by U.S. Department of Energy, Washington, D.C.

WMO (1979), *World Climate Conference*, WMO No. 537, World Meteorological Organization, Geneva.

WMO (1982), *The Stratosphere 1981. Theory and Measurements*, Global Ozone Research and Monitoring Project, Report No. 11, World Meteorological Organization, Geneva.

WMO (1985), *Atmospheric Ozone 1985*, Vols. I–III, Global Ozone Research and Monitoring Project, Report No. 16, World Meteorological Organization, Geneva.

2

Is Climate Stabilization
Still Feasible?

A. INTRODUCTION

Our review of past and recent temperature records showed that observed warming trends since the postwar period may still be within the limits of natural fluctuations. However, the full threat of climate change is not evident from such data. This is because of two factors:

- Past greenhouse gas emissions make their full impact felt only with significant delay, due to the inertia of the climate system.
- Continually growing GHG emissions cannot be eliminated instantaneously, due to the inertia of social and economic systems.

To determine whether the world's climate could still be stabilized at the levels discussed in Chapter 1, we pursue a two-part investigation: we first investigate the unrealized warming already "in the bank" from past emissions of trace gases, calculate the warming that can be expected under various scenarios of additional future emissions, and then determine how available emission reduction potentials could keep global warming within set limits.

The only tools available for these assessments are complex but limited computer models of the world's climate. For each scenario of future greenhouse gas emissions, these models can only predict a range of climate warming. The range of plausible emission scenarios is itself very wide, which adds to the uncertainty of the models.

To cope with this scenario diversity, the climate-model calculations of our study are designed to capture the widest possible range of proposed and practical emission scenarios. Using a high, medium, and low version for each GHG, we develop a "warming trumpet." This warming trumpet shows the range of warming to be expected, and can be directly compared to the climate-stabilization targets of Chapter 1.

We begin this chapter with a brief explanation of important technical aspects of climate modeling, including the overall modeling approach, the uncertainties inherent in various calibration and input data, and the distinction between ultimate warming commitment (equilibrium response calculations) and realized warming (transient response calculations). This review is followed by the scenario descriptions and warming results. The remainder of the chapter deals with the interpretation of these results for policy purposes.

B. CLIMATE-MODELING APPROACH

1. Concepts of Climate Modeling

The climate of the past can be assessed both by empirical studies based on observed data and by climate models. Assessment of the future climate response to radiative heating induced by greenhouse gases can only be done with mathematical models. These models are based on fundamental physical principles governing the climate system. Comprehensive climate response studies are conducted with interlinked submodels. The submodels describe the greenhouse response of the atmosphere, the thermal response of the oceans, and the diffusion and cycling of the various greenhouse gases in terrestrial, atmospheric, and ocean reservoirs.

Climate models come in various versions and degrees of complexity. The most sophisticated models are three-dimensional (3-d) general circulation models (GCMs), which allow some regional differentiation of climate change.

Equilibrium Response versus Transient Response. Climate modeling is used for two kinds of "experiments," referred to, respectively, as equilibrium[1] response studies and transient response studies (Tricot and Berger 1987). The difference between these concepts will be ex-

---

[1]Note the term *equilibrium* refers to thermal equilibrium.

plained, since it is important in interpreting warming results for policy decisions.

A change in the concentration of radiatively active trace gases results in a corresponding change in radiative forcing of the climate system. If there was no significant inertia in the climate system, as is the case for the atmosphere, a new equilibrium would be reached quickly. In reality, the heat capacity of the oceans, and to a lesser degree other factors such as continental ice shields, buffer against surface warming. Eventually, if the trace gas concentrations remain constant, an equilibrium is reached which is characterized by an *equilibrium global mean surface temperature T_e*. This temperature reflects the *total warming commitment* associated with greenhouse gas emissions.

At any point in time before equilibrium is reached between the oceans and the atmosphere, only a portion of the eventual warming is realized and thus observable in the temperature record. This *realized warming*, reflected in a *transient global mean surface temperature T_s*, is a function of the speed of heat absorption processes in the ocean.

Equilibrium response studies examine the climate response to a fixed radiative forcing (e.g., to a CO_2 doubling), or to a time-dependent radiative forcing (e.g., through "snapshot" experiments where successive equilibrium states are independently reached one after the other, based on a trace-gas emission scenario).

The equilibrium climate is related only to conditions in the atmosphere and perhaps to the upper ocean (mixed layer). Thus, in equilibrium response modeling, the complexities of the heat transfer processes within the deep ocean can be ignored. The surface-temperature change is calculated as though the climate system could respond instantaneously to trace-gas perturbations.

Whereas approximate equilibrium response studies can be computationally simple, calculations of transient response are more involved. In transient response studies, a coupled atmosphere–ocean model is needed to include the large heat capacity of the ocean and therefore allow determination of the lag in the response of the climate system. Again, the climate response is either calculated for an instantaneous, fixed radiative forcing (such as a CO_2 doubling), or for a scenario in which trace-gas concentrations change continuously over time. Calculating the transient response to a time-dependent, continuous change in CO_2 and other GHG is computationally much more demanding than equilibrium response analysis. Such studies have mainly relied on 1-d radiative–convective models of the atmosphere in conjunction with simple (box diffusion) ocean models. Recently, calculations using a 3-d general circulation model have been reported by Hansen et al. (1988; see Section D).

2. Climate Sensitivity and Uncertainty of Climate Estimates

The reliability of current climate-modeling estimates is still limited (Wigley 1984; WMO 1985; Bolin et al. 1986; Ramanathan 1988; Schneider 1989). It is therefore useful to identify some of the main areas of uncertainty.

Models are highly simplified replicates, both physically and mathematically, of the complex climate system, with differences in the kinds of simplifications used. In addition, historic data used to calibrate the models are not accurately known. For example, the observed global warming of 0.4–0.6°C over the past 100 years is uncertain largely because of incomplete data coverage and systematic errors, such as the urban heat island effect. Likewise, past trace-gas concentrations are uncertain because their sources, sinks, and chemical–climatic interactions are insufficiently understood.

There is also uncertainty regarding the warming–enhancing feedback mechanisms from cloud–radiation interactions, ocean and sea ice behavior, and land surface processes (including hydrology). Also a number of biogeochemical feedbacks are neglected in current models (Lashof 1989; see Section C in Chapter 1). To facilitate comparison among different models and model runs, a *net climate feedback factor f* is used, which is the ratio of the equilibrium surface warming to the warming that would have occurred in the absence of any feedbacks.

A significant degree of uncertainty in climate modeling is reflected in the wide range of equilibrium surface-temperature changes calculated for one and the same perturbation. The standard forcing used to compare models is a doubling of carbon dioxide concentration. An analysis by the U.S. National Academy of Sciences found a range of equilibrium temperature increase of $T_e = 1.5-4.5°C$ (CDAD 1983). Based on recent work with general circulation models, the uncertainty range may be 1.5–5.5°C (Dickinson 1986). If a fuller range of feedback mechanism is taken into account, sensitivities could be even higher (Lashof 1989).

As mentioned in Chapter 1, a range of climate sensitivities can also be calculated from greenhouse gas concentrations found in ice-core samples when combined with other paleoclimatic data. This yields a range of about 2–4°C for the climate sensitivity (Hansen 1989), approximately the same range as given by CDAC (1983). The possibility of major positive feedback effects from biogeochemical interactions, the higher climate sensitivities of the more sophisticated (3-d) general circulation models, and basic considerations of risk minimization all suggest that climate-stabilizing policies should be based on the upper end of the current uncertainty range in climate sensitivities.

The time-dependent calculations of the response of the climate system to greenhouse gas forcing are even more uncertain than the equilibrium ones. The time lag in the climate system before equilibrium temperature is reached is estimated to be anywhere between several decades and a century. This time lag is itself sensitive to other uncertain parameters, notably the equilibrium climate sensitivity, and the rate of mixing below the mixed ocean layer.

Uncertainty arises also from the incomplete knowledge of oceanic mixing processes. In the simple models presently used, these processes are parameterized by an eddy-diffusion coefficient. The mixing processes are, however, strictly diffusive and not necessarily well represented by such coefficients.

Nevertheless, considerable progress has been made in bringing modeled climate changes in accordance with observed data. An illustration of this predictive capacity, as well as some of the uncertainties, is shown in Figure 2.1, which is based on modeling work by Hansen et al. (1981).

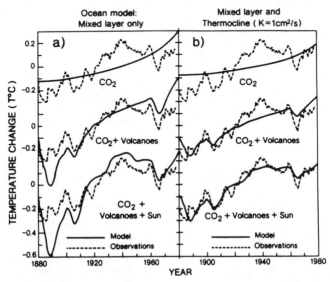

Figure 2.1 *Global temperature trend obtained from a global climate model for several assumed radiative forcings. Source: Hansen et al. (1981).*

3. Equilibrium Response Calculations

In our equilibrium response calculations, we use a scaling approach in which trace-gas concentrations are translated into temperature changes on the basis of simplified calculations of radiative forcing (see Appendixes 1 and 2).

Radiative Forcing. The radiative forcing of the atmosphere due to trace gases can be studied in two ways (Ramanathan et al. 1987). One approach is to look at the earth's surface and the troposphere as a system with a boundary at the tropopause, and to study the changes in the radiative energy fluxes in and out of this surface–troposphere system. The most sophisticated approach studies the changes in heating rates that are induced in each layer of the troposphere and captures changes in the vertical distribution of heating rates.

In the surface–troposphere system as a whole, the (mass-weighted) temperature change is governed by radiative forcing as defined by the change in net radiative flux at the tropopause boundary. The vertical distribution of temperature change within the troposphere is governed more by dynamical processes.

As a first approximation, we can ignore the details of vertical radiative forcing distribution and concentrate on the radiative forcing experienced by the surface–troposphere system as a whole. The calculation of trace-gas radiative forcing is then straightforward. The radiative forcing of an individual trace gas is assessed by fixing all other parameters such as temperature and humidity in the model and then calculating the changes in the radiative flux that are due only to a specific trace-gas concentration change. This approach is used in this book. Details are given in Appendix 1.

Since the greenhouse effects of the individual gases are largely additive, this method allows calculation of not only the contributions of individual gases, but also the total amount of equilibrium surface warming.

4. Transient Response Modeling

An overview of the data flow and model linkages is given in Figure 2.2. Estimated emission rates (historic or future) for carbon dioxide and chlorofluorocarbons are first processed through a carbon-cycle model and a CFC mass balance model to calculate the respective CO_2 and CFC concentrations. The resulting atmospheric CO_2 and CFC concentrations are then combined with estimates of historic and future CH_4

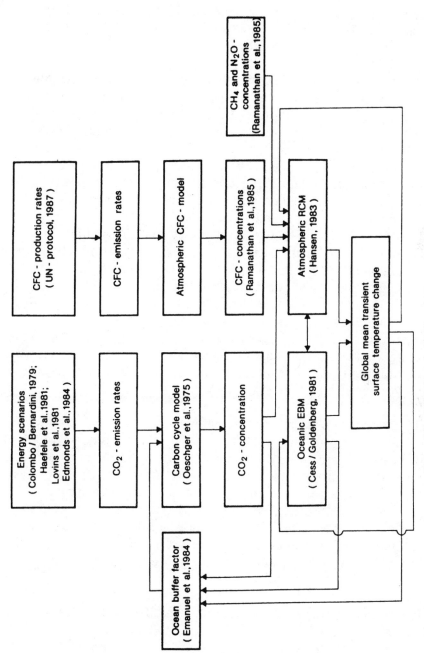

Figure 2.2 Modeling concept for transient response study.

and N_2O concentrations to serve as input in a climate model, which calculates the global mean surface temperature. The climate model consists of a (1-d) radiative–convective model (RCM) for the atmosphere and a (1-d) energy-balance model (EBM) for the ocean.

Carbon-Cycle Model. CO_2 emission rates are estimated according to various scenarios and serve as input to a carbon-cycle model, which calculates the historic and future atmospheric CO_2 concentrations. Net biospheric emissions are assumed to be zero. (For a detailed exploration of this assumption, see Section D in Chapter 3.) The box-diffusion carbon-cycle model developed by Oeschger et al. (1975) is used, which consists of three reservoirs acting both as sources and sinks, namely the atmosphere, a well-mixed ocean layer (about 75 m deep), and a deep sea (about 4000 m deep) subdivided into 42 layers in which transport is by eddy diffusion. The ocean chemistry that determines the efficiency of oceanic CO_2 uptake is calculated explicitly for each time step as a function of oceanic and atmospheric CO_2 concentration using the Oak Ridge carbon-cycle model (Emanuel et al. 1984). This approach allows computation of a time-dependent buffer factor, which is then used in the Oeschger et al. model (see also Fig. 2.2).

Other carbon-cycle models with different ocean dynamics and temperature feedback features are available. The range of carbon dioxide concentrations obtained from the major simple carbon-cycle models is further discussed in Chapter 4.

Mass Balance Model for CFCs. Base-year global CFC production rates are taken from Hammitt et al. (1986). Of the produced CFCs, only 70 percent are emitted into the atmosphere in the first year; the rest are released in exponentially decreasing quantities. The CFC mass balance in the troposphere and in the stratosphere is assumed to be 90:10. In the stratosphere, photolysis is described as the exponential decrease of an added CFC quantity. The concentrations by volume in the atmosphere are calculated from the molecular weights, the shares of the various masses, and the total mass of the atmosphere.

Climate Model. The climate model used here consists of the parameterized form of a 1-d radiative–convective model (RCM) of the atmosphere as developed by Hansen (1983), and a 1-d energy-balance model (EBM) of the ocean of a type similar to that used by Cess and Goldenberg (1981). With this model combination, the global mean transient surface-temperature change is estimated from 1860 to 2100. The RCM takes into account the concentration changes of CO_2, CH_4,

N_2O, CFC-11, and CFC-12, together with changes in surface temperature, volcanic activity, and the "solar constant." The latter two were ignored. Furthermore, neither the concentration changes in the other CFCs and O_3 nor the rather complex chemical reactions of the O_3 chemistry were considered. The CH_4 and N_2O expressions in the Hansen formulation were modified.

The energy-balance model (EBM) computes the heat flux into the ocean that results from the trace-gas concentration increases. For that purpose, the ocean is subdivided into 42 vertical layers—just as in the carbon-cycle model. The ocean-mixed layer (about 75 m) is assumed to be fully mixed. In the deep ocean (about 4000 m), energy transport is modeled as a vertical diffusion process. Of the many possible feedback processes, only the ocean buffer factor is explicitly taken into account (see Fig. 2.2). Other feedbacks are included in parameterized form. Further details are given by Hansen et al. (1981); Bach (1985); and Bach and Jung (1987).

C. HOW MUCH WARMING IS ALREADY LOCKED IN FROM THE PAST?

The question of the ultimate warming commitment from the past can be investigated by calculating the equilibrium response warming from past greenhouse gas emissions. Such an exercise also offers further insight into the quantitative aspects of the ocean-related time lag in climate warming.

1. Thermal Inertia of the Oceans

We consider, for the moment, only carbon dioxide, which accounts for most historic climate forcing from trace gases (see Figs. 2.7 and 2.8). The CO_2 increase from preindustrial levels of 275–290 ppm (1850) to 338 ppm (1980) can be translated into a range of equilibrium surface warmings for different climate sensitivities.

Figure 2.3 from Hansen et al. (1985) illustrates the results of such an exercise. It shows the surface warming ΔT_s and ΔT_e due to the CO_2 added to the atmosphere from 1850 to 1980. The warming results were calculated with a 1-d box-diffusion ocean model. The variables were the climate sensitivity ΔT_e (calculated for $2 \times CO_2$) and the net climate feedback factor f.

Consider, as an illustration, a climate sensitivity of 4.5°C for a doubling of CO_2. In this case, the equilibrium surface temperature

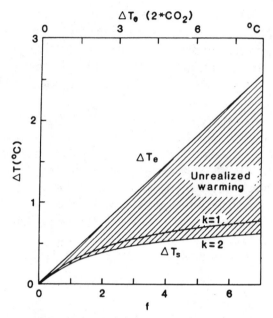

Figure 2.3 *Delay of the equilibrium warming by the ocean.*

increase ΔT_e from 1850–1980 CO_2 concentration changes, which were significantly less than doubling, is about 1.3°C. Reconstruction of the actual Northern Hemisphere surface-temperature history shows a warming of about 0.4–0.6°C from 1880 to 1980 (Jones et al. 1986). The discrepancy between these lower figures and the equilibrium figure is commonly interpreted to be the result of the thermal inertia of the ocean: the actual response of the climate system lags behind the equilibrium response that historic emissions will eventually produce.

Consider now the ocean effect. Hansen et al. (1985) show two curves based on calculation of observed ocean thermal diffusivities (κ) from tritium tracer studies. These measurements range between $\kappa = 1$–2 cm²/s. For the range of climate sensitivities from 1.5 to 4.5°C, the unrealized warming (shaded area in Fig. 2.3) is about half or more of the equilibrium warming. The 1850 to 1980 calculated observable (transient) warming ΔT_s in Figure 2.3 is only about 0.3 to 0.6°C, which agrees well with the observed temperature record.

These results lead to a number of important conclusions for policy-making purposes (WMO 1985; Hansen et al. 1985; Ramanathan et al. 1987; Ramanathan 1988):

- The present warming due to trace gases (the signal) is still relatively small and disguised by the natural variability of climate (the noise) because ocean heat storage makes ΔT_s about 45–75 percent less than ΔT_e.

- When global warming becomes evident in the world's monitoring network, a certain amount of continued warming will be inevitable even if no further emissions were to occur.

Insofar as the equilibrium warming measures the potential climatic change that we shall ultimately experience, it can be seen as our warming commitment (Mintzer 1987).

Figure 2.3 illustrates two further points:

- For a range of model climate sensitivities, the difference in transient warming is much smaller than the difference in expected equilibrium warming.
- The higher the climate sensitivity, the greater the portion of the warming commitment that remains hidden.

This feature of the climate system could help in a vigorous climate-stabilization effort in that it can buffer the present climate for a period of time from the full warming impact of past emissions. Moreover, this temporary buffer function grows disproportionately with increasing climate sensitivity. If the delay of the full warming is used as a grace period to turn around emission trends aggressively, realization of the full historic warming commitment could perhaps be avoided altogether, as discussed further below.

On the other hand, if no such action is taken, the delay in realizing the equilibrium warming could merely have the effect of reinforcing misperceptions about the magnitude of the risks from global warming. At the same time, higher climate sensitivities would lead to the accumulation of disproportionately large risks for future generations.

2. Limits to Determining Climate Risks from Monitoring Temperature Change

Figure 2.3 is revealing about the chances of soon narrowing the current range of model climate sensitivities on the basis of monitored warming. While the equilibrium temperature increases in a one-to-one ratio with model climate sensitivity, the transient response differs little over the range of model sensitivity, that is, the transient response curve is very shallow. This feature is of fundamental importance in the prevention versus wait-and-see issue: global warming, once it will have become evident in the temperature record, will be of limited help in settling the question of climate sensitivity.

Ironically, the chances for narrowing current uncertainties on the basis of monitoring global temperatures, while never great, are best in the worst possible outcome, that is, if a sizable warming materializes rapidly. Even then, many years of monitoring would be required to develop statistically relevant data. And while such monitoring goes on, significantly more warming commitment would be added. We conclude that

- The inertia of the climate system severely limits the degree to which temperature monitoring will be able to improve the decision-making basis.

3. Inertia in the Carbon Cycle

As pointed out in Chapter 1, the expected equilibrium warming is not necessarily fate. If emissions decrease enough over time to not only halt the rise in atmospheric concentrations but actually reduce them, radiative forcing is reduced and the full warming commitment from earlier emissions is not necessarily realized. In this sense, the inertia of the climate system could be beneficial: it could provide a chance to partially reverse earlier warming commitments before they are fully realized.

On the other hand, the degree to which concentrations of greenhouse gases can be lowered again is limited by a second kind of inertia in the climate system, that of the chemical cycles governing the greenhouse gases. This inertia is particularly large for carbon dioxide, which also happens to be the dominant driving force of climate change.

To illustrate the inertia-based "memory" of the combined chemical-cycle inertia and ocean thermal inertia, we have used our climate model to conduct a zero-emission sensitivity experiment for carbon dioxide. This experiment demonstrates how inertia exists (due to the ocean buffer effect), and is an important feature of the earth's natural carbon cycle (see also Section D in Chapter 3).

Figure 2.4 shows the CO_2 concentration from 1920 to 1980 calculated with the carbon-cycle model of Oeschger et al. (1975). The model parameters were fitted to the observed Mauna Loa data from 1960–1980. In the experiment, all fossil fuel use is stopped in 1980. As shown in the figure, the CO_2 concentration decreases by only 20 ppm from the 1980 value of 338 ppm to 318 ppm in 2100.

This decrease in the atmospheric CO_2 concentration over time (t)

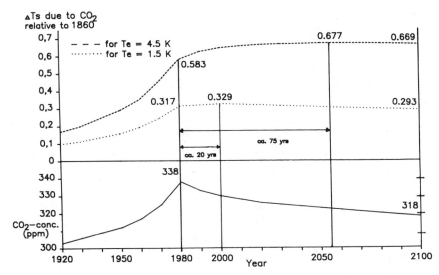

Figure 2.4 *Response in CO_2 concentration and global mean transient surface temperature (T_s) relative to 1860 for different climate sensitivities (T_e = 1.5–4.5 K), if all CO_2 emissions had been stopped in 1980.*

can be approximated by $CO_2(t) = CO_2(0) - [CO_2(0) - CO_2(1980)]$ $\exp(-t/100 \text{ yr})$, with $CO_2(0)$ being 309 ppm for 1920 and the "*e*-folding time" being 100 years. Once the earth's natural carbon cycle is perturbed, it will take approximately 600 years to fully return to an equilibrium state.

4. How Much of the Warming Commitment Will Inevitably Be Realized?

We next consider the transient warming response (ΔT_s) to a hypothetical complete halt in fossil fuel use in 1980. Again, Figure 2.4 shows the results. For the low climate sensitivity ($T_e = 1.5°C$), the warming continues to increase—albeit at a slow rate—for another 20 years, reaching about 0.329°C above the 1860 value. There is only a slow decline thereafter, and the ΔT_s is still significant (0.293°C) in 2100. Over the whole 120-year period, there is a reduction by 24 millidegree C below the 1980 temperature. Inspection of Figure 2.3 reveals that for the low climate sensitivity, virtually the entire warming commitment is still realized despite the emission cutoff. This is to be expected, since for low climate sensitivities the lag in the climate response is by definition small.

In the case of the higher climate sensitivity ($T_e = 4.5\,°C$), the temperature increases for another 75 years to peak at $\Delta T_s = 0.677\,°C$. The decline thereafter until 2100 is hardly measurable. Over the 120-year period, there is a net warming by 86 millidegree C. However, the maximum realized warming is only about half of the equilibrium warming predicted for that sensitivity (Fig. 2.3). Thus, while temperature continued to rise for a number of decades, the warming fate was substantially altered by emission curtailment.

This "Gedanken" (German for "thought") experiment leads to three conclusions of fundamental importance:

- Because of the unrealized warming contained in the memory of the climate system, a "wait-and-see" policy entails the risks of enormously increasing climate change. Depending on climate sensitivity, ultimate temperature changes could be more than twice the initially realized warming.
- Concentration reductions in trace gases could possibly avert the full realization of the equilibrium warming from past emissions, though temperatures would continue to rise for several decades under the momentum of past emissions.
- In order to bring about such concentration reductions, emissions of trace gases with long atmospheric residence times will have to be cut drastically.

These findings also clarify the relative roles of equilibrium and transient temperature changes in policy-making. So long as trace-gas concentrations, and therefore radiative forcing, do not decline, the equilibrium ΔT is what measures the potential climatic change that we shall ultimately experience. Risk assessments for policy purposes should take into account the equilibrium warming, and not just the lower transient warming.

Have Emissions to Date Already Preempted Climate Stabilization? In determining whether our past actions have already preempted climate stabilization as defined in Chapter 1, we need to consider future emissions as well, including answers to these three questions:

- What is the expected equilibrium warming from past emissions over the range of plausible climate sensitivities?
- What further increments in warming commitment will be added

over the next few decades when concentration reductions will be difficult or impossible to achieve?

- And what scenarios, if any, could prevent the full realization of both current and further warming commitments over the next century?

The latter two questions are addressed in Section D, where we first calculate the additional warming increment from a range of published energy scenarios and other trace-gas projections until 2030. We then use transient response modeling to calculate how much of this warming commitment would be expected to materialize until 2100, given alternative emission trajectories.

For the moment, we examine the expected equilibrium warming from emissions up to the present. Figure 2.3 indicates that the equilibrium warming obtained for a particular climate sensitivity can be linearly scaled to calculate the warming at other sensitivities. Over the climate-sensitivity range of 1.5–5.5°C, the warming commitment locked in by 1850–1980 releases of carbon dioxide alone remains at about 0.5–1.8°C relative to 1850—significantly below the limit of 2.5°C. In section D.1, we show what results are obtained from including the other trace gases as well. At a climate sensitivity of 2.3°C, the range of estimates for emissions until 1980 is 0.74–0.94°C. Scaling these figures to a climate sensitivity of 5.5°C extends this range all the way to 2.4°C. This range is close to the estimates by Ramanathan et al. (1987). Thus,

- If climate sensitivity is 5.5°C or greater, climate stabilization as defined in this text could only be achieved through reductions in radiative forcing below present levels.

Based on the same climate sensitivities, the world is currently adding a warming commitment of 0.13–0.5°C per decade (Ramanathan 1988). If realistic at all, the chance of reducing total radiative forcing to below present levels before the equilibrium warming from present levels is fully realized will become progressively more elusive as greater and greater additional commitments are made.

D. WHAT FURTHER WARMING WILL FUTURE EMISSIONS BRING?

In this section, we investigate a range of scenarios for greenhouse gas emissions. We calculate the ultimate warming commitment implied in each scenario (equilibrium response), spanning the time from 1980 until 2030 and using the 1860–1980 period as a historic basis.

We also calculate expected transient surface-temperature changes that will be observable. To capture the lingering warming effects that will still be felt after effective greenhouse gas curtailment has been initiated or achieved, we use an extended time horizon to the year 2100.

1. Emission Scenarios

Energy Scenarios. Future world energy use is one of the major uncertainties in predicting climate change. To allow for this, the usual approach in energy analysis is to investigate a range of scenarios. We use the highest and lowest published energy scenarios, respectively, to set the lower and upper limit of fossil fuel impacts. The highest published scenario is referred to as "Oak Ridge A" (Edmonds et al. 1984). The lowest published scenario is referred to as "Energy Efficiency" (Lovins et al. 1981). As a medium scenario, we use the "IIASA Low" scenario (Haefele et al. 1981).

Table 2.1 gives detailed information on changes in fossil fuel use by source, on C emissions, and on CO_2 concentration for the high (Oak Ridge A) and low (Energy Efficiency) scenario. "Oak Ridge A," as developed by Edmonds et al. (1984), projects fossil fuel demand to 2075. We extrapolated it to 2100, using trends in the period after 2050 as a basis. In this scenario, coal overtakes oil to become the dominant fuel as early as 1990. Total annual fossil fuel use increases from 7.83 Terawatt-years per year (TW-yr/yr) in 1980 to about 159 TW yrs by 2100. The enormous fossil fuel use would increase fossil carbon releases 22-fold, from the 1980 value of 5 Gt/yr to about 111 Gt/yr in 2100. Such levels are clearly not plausible. Note, however, that this result is based on a growth rate of only 2.5 percent per year.

This leads to a staggering cumulative fossil fuel use of some 7450 TW-yr by the end of the next century. At that growth rate, the economically recoverable global fossil fuel reserves of about 1000 TW-yr would be exhausted by approximately 2030. By 2100, 80 percent of all estimated conventional coal, oil, and gas resources would be consumed (see Chapter 4). Thus, the resource constraints alone, and more so the economic and environmental costs, make this an unlikely scenario much beyond 2030.

A stark contrast to this picture is provided by the "Efficiency" scenario developed by Lovins et al. (1981). It depicts a least-cost strategy in which all the presently available cost-effective efficiency

TABLE 2.1   Energy Use, C Emissions, and CO_2 Concentration for the Highest and the Lowest Scenarios

Energy scenario	No.	year	Fossil fuel use (TW) oil	gas	coal	total annual	cumulative from 1980	reserves to be exhausted in year	C-Emission (Gt) total annual	cumulative from 1980	CO_2 conc. ppm
		1900	0.02	0.01	0.63	0.66			0.5		293
		1980	3.70	1.69	2.44	7.83	7.83		4.96	4.96	338
		1990	4.31	2.40	4.38	11.09	104.03		7.13	66.49	360
Oak-Ridge A		2000	4.92	3.11	6.31	14.34	232.79		9.30	149.74	386
		2010	6.39	4.03	10.39	20.82	411.80		13.76	267.28	425
(Edmonds		2020	7.86	4.96	14.48	27.29	655.58		18.21	429.35	479
et al., 1984)	1	2030	10.88	5.34	20.14	36.36	971.90	2026*	24.63	641.87	551
		2040	15.46	5.17	27.39	48.02	1399.63	or	33.03	934.37	656
		2050	20.04	5.01	34.63	59.68	1943.96	2065**	41.42	1310.77	794
		2060	30.81	4.35	44.27	79.43	2649.37		55.39	1801.77	982
		2070	41.58	3.69	53.91	99.18	3552.26		69.36	2432.45	1234
		2080	52.34	3.08	63.55	118.98	4652.80		83.35	3202.89	1550
		2090	63.11	2.58	73.19	138.89	5952		97.39	4113.54	1932
		2100	73.88	2.17	82.83	158.88	7450.75		111.45	5164.77	2474
		1900	0.02	0.01	0.63	0.66			0.5		293
		1980	3.70	1.69	2.44	7.83	7.83		4.96	4.96	338
		1990	3.15	1.73	2.40	7.27	87.22		4.59	55.21	357
Efficiency		2000	1.77	1.52	1.77	5.05	147.72		3.13	93.08	377
		2010	1.26	1.12	1.31	3.69	190.72		2.29	119.75	380
		2020	0.75	0.73	0.84	2.32	220.09	will	1.44	137.95	382
(Lovins et al.,	5	2030	0.24	0.34	0.38	0.96	235.82	not be	0.59	147.66	381
1981/83)		2040	0.10	0.17	0.19	0.46	242.54	reached	0.28	151.78	378
		2050	0.04	0.09	0.09	0.21	245.61		0.13	153.65	375
		2060	0.01	0.04	0.04	0.10	247.03		0.06	154.51	373
		2070	<0.01	0.02	0.02	0.05	247.69		0.03	154.91	370
		2080		0.01	0.01	0.02	248.01		0.01	155.10	368
		2090		<0.01	<0.01	0.01	248.16		<0.01	155.19	366
		2100				0.01	248.23			155.23	364

Economically recoverable fossil fuel reserves are ca. 836 TW and those probably technically recoverable are ca. 3084 TW** (Ziegler and Holighaus, 1979). 1 Terawatt-years per year = 1 TW.

improvements in the use of energy are more or less fully implemented over the next 50 years. (See Chapter 6 for a more detailed discussion.) In this scenario, world energy use declines to about half of current levels by 2030, and fossil fuel use declines to about 12 percent of 1985 levels. For the purpose of this exercise, we extrapolate this declining trend to 2100, which leads to essentially zero fossil carbon emissions in the period after 2060 (see Table 2.1).[2]

The average rate of decline in the consumption of fossil fuels is approximately 3.2 percent per year. The scenario initially assumes a

[2]Note that releases of carbon dioxide do not have to decline all the way to zero to stop the buildup of carbon dioxide in the atmosphere. As shown in Table 2.1, the atmospheric CO_2 concentration actually declines after about 2030 in the efficiency scenario. For further discussion of this point, see Section D in Chapter 3, and Chapter 4.

slower rate of decline to reflect implementation barriers, and a faster reduction rate later on when policies to promote efficiency investments have picked up momentum. The cumulative fossil fuel use never gets close to exhausting even current economically recoverable reserves. The cumulative C-release is only $\frac{1}{33}$ of that in the "Oak Ridge A" scenario.

A widely quoted middle-of-the-road scenario is the "IIASA Low" scenario developed by Haefele et al. (1981).[3] Details are not given here in tabular form. In this scenario, energy demand projections cover the years from 1975–2030. For the purpose of our exercise, we use a linear extrapolation to 2100 on the basis of the growth rate in the 2020–2030 period. The resulting cumulative fossil fuel use is about 1950 TW-yr in 2100, exhausting the presently economically recoverable fossil fuel reserves by about 2050. By 2100, the estimated cumulative C-emissions are eight times higher than those obtained for the "Efficiency" scenario.

CFC Scenarios. Of all the trace gases, the CFCs are presently closest to being put under some kind of worldwide control. Forty-three U.N. member countries have signed the so-called U.N. or Montreal Protocol on the control of chlorofluorocarbons and halons. The principal motivation behind these proposed control measures is the protection of the stratospheric ozone layer. The emphasis in our analysis is, however, on the contribution of CFCs to global warming. A summary of the climatic and ozone-depleting effects of important chemicals is given in Table 3.3. Chapter 3 gives details on the Montreal Protocol and on control options. There is not enough information available at this time to reliably estimate the past and future contribution to warming from all substances covered under the protocol.

The scenarios for the CFCs are again selected to bracket possible future emission rates. Base-year CFC production rates are taken from Hammitt et al. (1986) (see Section B in Chapter 3). The high, medium, and low scenarios from 1980 to 2030 as used in the equilibrium response calculations are shown in Tables 2.2 and 2.3, based on projections given by Ramanathan et al. (1985). Alternatively, we use a scenario based on the proposed CFC reduction rates of the Montreal Protocol, assuming 100 percent participation by the nations of the world. Note, however, that as agreed upon, the protocol provided a

---

[3]For a methodological critique of this study, see, for example, Keepin and Wynne (1984).

TABLE 2.2 Trace-Gas Scenarios for Equilibrium Temperature Change Calculations

Scenario				Concentration[1] 1980	2030	Future growth rate[2] %/yr
HIGH	A	Oak Ridge A[3] Ramanathan et al. (1985)	for CO_2	338	551	0.98
			for CH_4	1650	3300	1.40
			for N_2O	300	450	0.81
			for CFC-11	0.179	2	4.95
			for CF-12	0.307	3.5	4.99
	B	for CO_2, CH_4, N_2O same as A UN-Protocol[4]	for CFC-11	0.179	0.318	1.16
			for CFC-12	0.307	0.514	1.04
MEDIUM	C	Ramanathan et al. (1985)	for CO_2	338	450	0.57
			for CH_4	1650	2340	0.70
			for N_2O	300	375	0.45
			for CFC-11	0.179	1.051	3.60
			for CF-12	0.307	3.5	4.99
	D	for CO_2, CH_4, N_2O same as C UN-Protocol same as B				
LOW	E	Energy Efficiency[5] Ramanathan et al. (1985)	for CO_2	338	381	0.24
			for CH_4	1650	1850	0.23
			for N_2O	300	350	0.31
			for CFC-11	0.179	0.5	2.08
			for CF-12	0.307	0.9	2.17
	F	for CO_2, CH_4, N_2O same as E UN-Protocol same as B				

		Current growth rate %/yr
for CO_2	(Gammon et al. (1985) Wigley, 1987)	0.4
for CH_4	(Rasmussen and Khalil, 1986)	1.1
for N_2O	(Weiss, 1981; Dickinson and Cicerone, 1986)	0.3
for CFC-11	(Khalil and Rasmussen,	5.0
for CFC-12	1981; Fabian, 1986)	5.0

[1] CO_2 is in ppm; all others are in ppb.
[2] Concentration changes for Oak Ridge A and Energy Efficiency from scenario analysis; all others are exponential projections.
[3] Edmonds et al. (1984).
[4] Based on a 1990 freeze of CFC-11 and CFC-12 production rates at 1986 levels, a 20% cut in 1994 and another 30% cut by 2000; calculations made with a mass balance model.
[5] Lovins et al. (1981).

number of legal exemptions for both industrialized and developing countries, and full Third World participation is not ensured. Only if these exemptions are ignored and if full participation is assumed would the protocol result in the emission reductions assumed in our Montreal Protocol scenarios (see also Section B in Chapter 3).

TABLE 2.3 Additional Chlorofluorocarbons Considered in This Text

Compound	Main use	Estimated average life-time yrs.	Global average atm.conc. (ppt)[1] 1980	Growth rates (%/yr) 1980–2030		
				High	Medium	Low
CH_3CCl_3	Degreasing solvent	8	140	6.77	4.86	3.27
CFC-22 ($CHClF_2$)	Refrigerant, sterilant	20	60	7.15	5.56	3.87
CCl_4	Used for CFC-production	50	130	2.27	1.69	0.87
CFC-14 (CF_4)	Aluminum industry	500	70	3.02	2.49	2.12
CH_2Cl_2	Solvent	0.6	30	4.71	3.87	2.44
CFC-13 ($CClF_3$)		400	7	5.46	4.39	3.55
$CHCl_3$	Used for CFC-22 production	0.7	10	4.71	2.22	1.39
CFC-116 (C_2F_6)	Aluminum industry	500	4	4.71	3.27	1.85

1. ppt = parts per trillion = 10^{-12}
Extracted from Ramanathan et al. (1985)

Source: Ramanathan et al. (1985).

There are other CFCs not covered by the Montreal Protocol that eventually may contribute to global warming. Eight such substances are listed in Table 2.3, and their warming impact is incorporated in the equilibrium response calculations as well. Overall, the CFC scenario incorporates contributions of 10 CFCs to warming.

For the transient response analysis, we use only one CFC scenario, which is based on extrapolating the Montreal Protocol limits so as to yield approximately constant concentrations of CFC-11 and CFC-12 by the year 2100.

Recent research, including a report by the U.S. Environmental Protection Agency, suggests that a complete phaseout of CFCs by the turn of the century may be needed to preserve the stratospheric ozone layer (Hoffman and Gibbs 1988). If such plans materialize, the Montreal Protocol could come to represent the upper end of the range of likely future CFC concentrations. We examine the impact of a complete and early phaseout on our warming results in Sections D.2 and D.3 below.

Nitrous Oxide and Methane. We do not use emission scenarios but concentration scenarios for these gases. Here, again, high, medium, and low scenarios up to 2030 are based on Ramanathan et al. (1985). For the transient response calculations, only a medium and a low scenario are used. In the medium case, the 2020–2030 concentration changes of the "best estimate" by Ramanathan et al. are linearly extrapolated to 2100. CH_4 concentrations in 2100 are 3.31 ppm, or about twice the 1980 level. N_2O concentrations rise by 50 percent to 0.480 ppm.

In the low scenario, the CH_4 concentration asymptotically stabilizes in 2100 at 10 percent above 1980 levels, or 1.81 ppm. The same holds for the N_2O concentration, which reaches 0.328 ppm in 2100. Maintaining approximately constant levels of concentration would require a cut in present emissions of about 10–20 percent for methane, and about 80–85 percent for nitrous oxide (Lashof and Tirpak 1989).

Combined Emission Scenarios. For the equilibrium response calculations, we examine six sensitivity cases (Table 2.2). These form two series of three: a high (A), medium (C), and low (E) case for energy, CFCs, N_2O, and CH_4, where we use the CFC projections by Ramanathan et al. (1985); and a high (B), medium (D), and low (F) case, where we use the Montreal Protocol to modify CFC emissions. Current concentration growth rates and the various assumptions on future growth rates are summarized in Tables 2.2 and 2.3.

For the transient response studies, we combine the high and medium energy scenarios with the medium scenarios for CFCs, N_2O, and CH_4 based on Ramanathan et al. (1985). For the low energy scenario, we use the extrapolation of the Montreal Protocol and the low scenario for N_2O and CH_4.

2. Equilibrium Response Studies

Trace-Gas Concentrations. Figure 2.5a–d and Table 2.4 show the calculated trace-gas concentration increases from 1960 to 1980, and from 1980 to 2030, for the range of scenarios given in Tables 2.2 and 2.3. The concentrations up to 1980 are based on empirical expressions deduced by Wigley (1987) from observations. The projections to 2030 are fitted to growth rates given by Ramanathan et al. (1985). The resulting formulas for the respective gases and time periods can be found in Appendix 1.

For CO_2, Figure 2.5a shows an increase of 13 percent, 33 percent, and 63 percent in 2030 over 1980 for the low, medium, and high scenarios, respectively. Note that the low ("Efficiency") scenario begins to show a stabilization in CO_2 concentration by the late 2020s.

Large concentration increases also result for the other trace gases. Compared to 1980, the 2030 concentrations for CH_4 would be 12 to • 100 percent higher (Fig. 2.5b), and those for N_2O would be 16–49 percent higher (Fig. 2.5c). The CFC projections show even greater changes, reaching 2.8 to 11.4 times the 1980 figure (Fig. 2.5d). In the

72

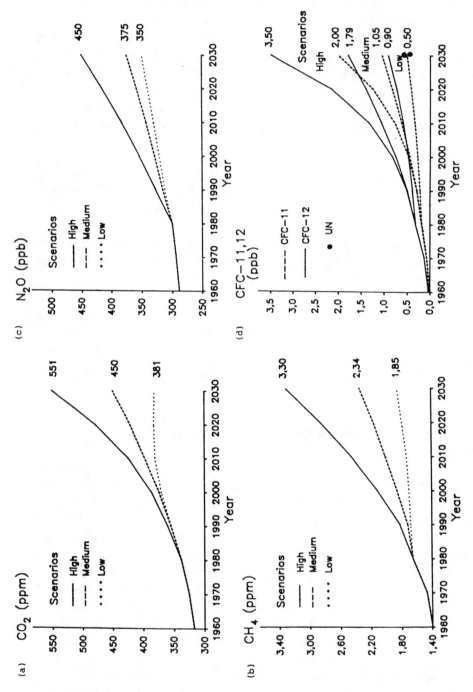

Figure 2.5 Calculated trace-gas concentration change from 1960–1980, and for the high, medium, and low scenarios given in Table 2.2 from 1980–2030—for (a) CO_2, (b) CH_4, (c) N_2O, and (d) CFC-11 and CFC-12.

TABLE 2.4 Historic and Projected Atmospheric Concentrations of Important Trace Gases for a Range of Scenarios

Year	Scenario[1]	Atmospheric Concentrations					Equivalent CO_2 concentration ppm[2]
		CO_2	CH_4	N_2O	CFC-11	CFC-12	
		ppm	ppm	ppm	ppb	ppb	
1850		287.4	0.863				287.4
1900		296	0.974	0.280			299
1950		312	1.136	0.285			321
1980		338	1.650	0.303	0.179	0.307	378
1990	A	360	1.896	0.327	0.290	0.499	422
	B	360	1.896	0.327	0.219	0.356	417
	C	355	1.769	0.315	0.286	0.486	410
	D	355	1.769	0.315	0.219	0.356	405
	E	354	1.688	0.312	0.220	0.380	401
	F	354	1.688	0.312	0.219	0.356	401
2000	A	386	2.178	0.354	0.470	0.813	481
	B	386	2.178	0.354	0.252	0.400	464
	C	374	1.897	0.328	0.438	0.745	450
	D	374	1.897	0.328	0.252	0.400	436
	E	377	1.727	0.321	0.270	0.472	434
	F	377	1.727	0.321	0.252	0.400	433
2010	A	425	2.503	0.383	0.762	1.323	575
	B	425	2.503	0.383	0.274	0.438	530
	C	397	2.034	0.342	0.616	1.047	500
	D	397	2.034	0.342	0.274	0.438	472
	E	380	1.767	0.330	0.332	0.584	445
	F	380	1.767	0.330	0.274	0.438	441
2020	A	479	2.877	0.415	1.236	2.153	727
	B	479	2.877	0.415	0.296	0.476	622
	C	422	2.187	0.358	0.820	1.394	559
	D	422	2.187	0.358	0.296	0.476	512
	E	382	1.808	0.340	0.408	0.724	457
	F	382	1.808	0.340	0.296	0.476	448
2030	A	551	3.300	0.450	2.000	3.500	986
	B	551	3.300	0.450	0.318	0.514	746
	C	450	2.340	0.375	1.051	1.787	630
	D	450	2.340	0.375	0.318	0.514	558
	E	381	1.850	0.350	0.500	0.900	466
	F	381	1.850	0.350	0.318	0.514	452

[1]For explanation of scenarios, see Tables 2.2 and 2.3.
[2]Including the CFCs shown in Table 2.3.

ideal but not very realistic case in which the legally permitted exemptions of the Montreal Protocol were disregarded, the concentration increase would be limited to about 70 percent of 1980 levels.

Equivalent CO_2 Concentrations and Global Warming Commitment. The individual concentrations of each trace gas can be converted into an equivalent CO_2 concentration. This index is the total concentration of all gases expressed in terms of a CO_2 equivalent that would produce the same amount of radiative forcing, and therefore warming. (See Appendix 2 for the method used in this conversion.)

Figure 2.6 shows the estimated global concentration and warming trends for the high, medium, and low scenarios (inputs from Tables 2.2 and 2.3). CO_2, CH_4, N_2O, and the ten CFCs are included. The plot covers the historic period from 1850 to 1980, and the scenario period from 1980 to 2030. The changes are given both in terms of equivalent CO_2 concentrations (left ordinate; see also Table 2.4), and in terms of the global average equilibrium surface temperature (right ordinate). The warming is based on a climate sensitivity of 2.3°C (see Appendix 2).

Figure 2.6 and Table 2.4 illustrate a very important finding. While CO_2 concentrations will be a little more than 350 ppm by 1990, the equivalent CO_2 concentration in the same year will be already about 400 ppm. Likewise, if CO_2 from fossil fuel combustion was the only source of greenhouse gases, a doubling of CO_2 from the preindustrial value of 290 ppm to 580 ppm would not occur before the second half of the next century under any of the scenarios. By contrast, a doubling on an equivalent CO_2 basis would already occur by 2011 in the high scenario A. By 2030, the doubling mark based on CO_2 equivalence has been passed in scenarios A, B, and C. This means that levels of climate risk formerly associated with the doubling of CO_2 are reached much sooner once the other trace gases are taken into account.

Overall, there are major differences in the expected global warming (Fig. 2.6 and Table 2.5). For a climate sensitivity of 2.3°C, total warming in 2030 relative to 1850 ranges from 1.5 to 4.2°C.

Relative Contribution of Individual Trace Gases to Warming. In assessing the relative importance of control policies in different areas, it is necessary to know how much warming is contributed by each greenhouse gas. To put things into context, we compare historic (1850–1980) contributions to equilibrium surface warming with future (1980–2030) warming contributions. This information is given in Table 2.5, which allows comparison of the contributions between individual gases, time periods, and scenarios. Cumulative results for the two periods are also graphically shown in Figures 2.7 and 2.8. For each greenhouse gas, the contribution to total warming commitment can be read off the graphs both in terms of degrees (relative to 1850) and as a percentage share.

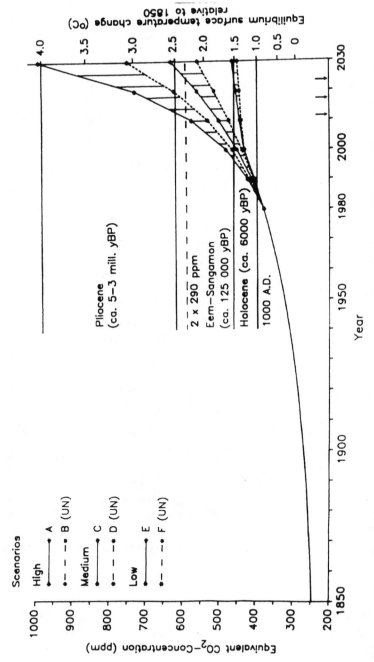

Figure 2.6 Global equilibrium warming due to trace gases, 1850–2030, and the effect of the Montreal Protocol.

75

TABLE 2.5 Contribution of Trace Gases to the Past and Future Greenhouse Effect for a Range of Scenarios

Year	Scenario [1]	Equilibrium Temperature Change (CO_2) [2] relative to 1850							
		CO_2	CH_4	N_2O	CFC-11	CFC-12	Other CFC	All CFC	All Gases
1850		0.000	0.000	0.000	0.000	0.000	0.000	0.000	0.000
1900		0.097	0.038	0.000	0.000	0.000	0.000	0.000	0.135
1950		0.270	0.089	0.009	0.000	0.000	0.000	0.000	0.368
1980		0.534	0.231	0.037	0.025	0.049	0.023	0.098	0.900
1990	A	0.742	0.291	0.074	0.041	0.080	0.037	0.159	1.265
	B	0.742	0.291	0.074	0.031	0.057	0.032	0.120	1.226
	C	0.696	0.261	0.055	0.040	0.078	0.038	0.156	1.168
	D	0.696	0.261	0.055	0.031	0.057	0.032	0.120	1.131
	E	0.686	0.240	0.051	0.031	0.061	0.028	0.120	1.098
	F	0.686	0.240	0.051	0.031	0.057	0.032	0.120	1.097
2000	A	0.971	0.355	0.114	0.066	0.131	0.060	0.257	1.697
	B	0.971	0.355	0.114	0.035	0.064	0.038	0.138	1.578
	C	0.867	0.291	0.075	0.061	0.120	0.058	0.240	1.473
	D	0.867	0.291	0.075	0.035	0.064	0.038	0.138	1.372
	E	0.894	0.250	0.065	0.038	0.076	0.034	0.148	1.356
	F	0.894	0.250	0.065	0.035	0.064	0.038	0.138	1.346
2010	A	1.288	0.424	0.155	0.107	0.213	0.097	0.417	2.284
	B	1.288	0.424	0.155	0.038	0.071	0.041	0.150	2.017
	C	1.064	0.323	0.096	0.086	0.169	0.082	0.337	1.820
	D	1.064	0.323	0.096	0.038	0.071	0.041	0.150	1.633
	E	0.920	0.260	0.078	0.047	0.094	0.041	0.182	1.440
	F	0.920	0.260	0.078	0.038	0.071	0.041	0.150	1.408
2020	A	1.682	0.498	0.199	0.174	0.347	0.156	0.676	3.055
	B	1.682	0.498	0.199	0.042	0.077	0.044	0.162	2.541
	C	1.265	0.357	0.119	0.115	0.225	0.109	0.449	2.190
	D	1.265	0.357	0.119	0.042	0.077	0.044	0.162	1.904
	E	0.937	0.270	0.093	0.057	0.117	0.049	0.223	1.523
	F	0.937	0.270	0.093	0.042	0.077	0.044	0.162	1.462
2030	A	2.143	0.577	0.245	0.281	0.564	0.249	1.094	4.059
	B	2.143	0.577	0.245	0.045	0.083	0.046	0.174	3.138
	C	1.477	0.390	0.144	0.148	0.288	0.139	0.575	2.585
	D	1.477	0.390	0.144	0.045	0.083	0.046	0.174	2.184
	E	0.928	0.280	0.108	0.070	0.145	0.058	0.274	1.590
	F	0.928	0.280	0.108	0.045	0.083	0.046	0.174	1.49

[1] For explanation of scenarios, see Table 2.4.
[2] Calculations are for a climate sensitivity of 2.3°C for a doubling of CO_2 (see Appendix 1 for details).

Carbon Dioxide. Between 1860 and 1980, the dominant contribution was from CO_2, which had a share of 59 percent of the historic warming commitment. For the period 1980–2030, the relative share of CO_2 is somewhat reduced if CFC control is not implemented, but at about 51–57 percent (Fig. 2.7), CO_2 is still the dominant driving force of warming. With the Montreal Protocol implemented, there would be

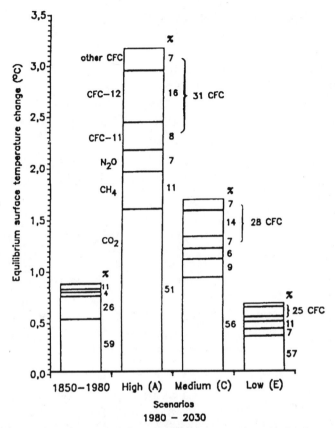

Figure 2.7 *Historic and future contribution of trace gases to global warming for a range of scenarios without CFC control.*

somewhat of an increase, to 67–73 percent. These figures underline the importance of reductions in fossil fuel use.

In absolute terms, the contribution to temperature change from CO_2 was 0.53°C for the 130 years from 1850 to 1980. It is expected to reach 1.6°C in the high scenario over a period of only 50 years. This would be a dramatic, eight-fold acceleration of the warming commitment, from an average of 0.04°C per decade to 0.32°C per decade. By contrast, the rate of warming commitment in the low scenario would be close to historic levels. The average CO_2 contribution would be only 0.39°C from 1980 to 2030, or 0.08°C per decade.

As can be seen from Table 2.5, the fossil fuel reduction measures in the low scenarios E and F begin to show an effect as early as 2010, with the absolute warming commitment from CO_2 leveling off at 0.9°C first observed in measurable amounts in the early 1960s. In the future, their share depends strongly on the measures taken to protect the ozone layer.

In absolute terms, the contribution to global warming of all CFCs was about 0.1°C for 1960–1980; it could increase ten-fold without Montreal control in the high scenario A. Under full implementation of the protocol without use of permitted exemptions, it would double. In percentage terms, the Montreal Protocol has the following impacts:

- Without the protocol, the warming contribution of the CFCs would roughly treble to 25–32 percent, making them the second largest driving force of global warming after CO_2 (Fig. 2.7);
- With full implementation of the protocol and no use of exemptions, the CFC contribution to warming shrinks to 3–13 percent (Fig. 2.8).

These findings show that international agreements like the Montreal Protocol can be a very important factor in limiting warming.[4] The same figures also show that fossil fuel reductions are the overriding factor in preventing a CO_2-equivalent doubling. This underscores energy policy's preeminent role in climate stabilization.

Contribution to Warming from Individual CFCs. Figure 2.9 shows the contribution of individual CFCs to the total warming commitment from all CFCs. The figure shows the shift in percentage shares relative to 1960 for three cumulated time periods. The 2030 warming commitment from CFC-12 is almost double that of CFC-11. CFC-12 and CFC-11 are presently the dominant greenhouse gases among the CFCs, with a share of 76 percent.

The impact of the Montreal Protocol is also shown (shaded area). If applied only to CFC-11 and CFC-12, the protocol would reduce the warming commitment from all ten CFCs to 44 percent of the medium trend projection (scenario C). The warming commitment from CFC-11 and CFC-12 alone would be reduced to 31 percent. If the protocol were applied to the whole group of ten CFCs considered here,[5] the warming would be reduced to 23 percent of the no-control case.

---

[4]Because of the permitted exemptions, the impact of the protocol will be diluted. It is now widely recognized that the protocol will need to be strengthened.

[5]It is assumed that the U.N. protocol is only applied to CFC-11 and CFC-12, while the other eight CFCs grow according to the "medium" projections shown in Table 2.3. On that basis, total warming from all CFCs is 0.18°C, as shown in Table 2.5. The case of applying the protocol to the other eight CFCs is not represented among the six scenarios in Table 2.5.

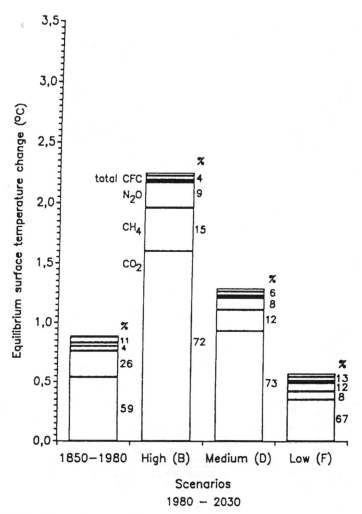

Figure 2.8 *Historic and future contribution of trace gases to global warming for a range of scenarios with Montreal CFC control.*

around 2030 and then beginning a very slow downward trend in 2030 and beyond (not shown here).

CH₄ and N₂O. These gases contributed 26 percent and 4 percent, respectively, to the historic warming commitment. Over the period 1980–2030, the N_2O share is expected to double, while CH_4 would lose more than half its share in all three scenarios.

Chlorofluorocarbons. By 1980, the total CFC contribution was already a surprising 10 percent, considering that CFCs in the atmosphere were

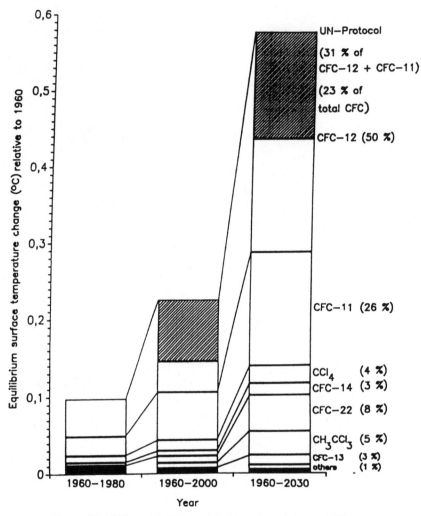

Figure 2.9   CFC contribution to global warming relative to 1960.

Comparison with Other Studies. To assess the relationship of these model calculations to other studies, we compare our results with those obtained by Ramanathan et al. (1985) and Tricot and Berger (1987), using three different time periods (Table 2.6). All results are based on a climate sensitivity of about 2°C.

The estimated historic temperature increase from all gases combined is similar in all studies, ranging from 0.74 to 0.90°C over the past 100–130 years. The estimates for the contributions of all CFCs are also similar. Our study shows higher estimates for CH_4 and N_2O.

TABLE 2.6 Comparison of Historic and Future Trace-Gas Contributions to Global Equilibrium Warming as Obtained by Different Authors and Scenarios

Period	Authors/Scenarios	Equilibrium Surface Temperature Change (°C)														
		CO_2 °C	%	CH_4 °C	%	N_2O °C	%	CFC-11 °C	%	CFC-12 °C	%	Other CFC[4] °C	%	All CFC °C	%	All Gases °C
1850/80 –1980	Ramanathan et al. (1985)	0.52	70	0.12	16	0.02	3	0.02	3	0.04	5	0.02	2	0.09	11	0.74
	Tricot & Berger (1987)[1]	0.61	74	0.11	14	0.02	2							0.07	10	0.82
	This study[2]	0.53	59	0.23	26	0.04	4	0.03	3	0.05	6	0.02	3	0.10	11	0.90
1980–2030	Ramanathan et al.	0.71	52	0.14	10	0.10	7							0.43	31	1.38
	Tricot & Berger	0.80	66	0.19	16	0.06	5							0.16	13	1.21
	This study Scenarios[3] A	1.61	51	0.35	11	0.21	7	0.26	8	0.52	16	0.23	7	1.00	32	3.16
	B	1.61	72	0.35	15	0.21	9	0.02	1	0.03	2	0.02	1	0.08	3	2.24
	C	0.94	56	0.16	09	0.11	6	0.12	7	0.24	14	0.12	7	0.48	28	1.69
	D	0.94	73	0.16	12	0.11	8	0.02	2	0.03	3	0.02	2	0.08	6	1.29
	E	0.39	57	0.05	07	0.07	10	0.05	7	0.10	14	0.04	5	0.18	25	0.69
	F	0.39	67	0.05	08	0.07	12	0.02	3	0.03	6	0.02	4	0.08	13	0.59
1850/80 –2030	Ramanathan et al.	1.23	58	0.26	12	0.12	6							0.51	24	2.12
	Tricot & Berger	1.41	69	0.30	15	0.08	4							0.23	12	2.03
	This study A	2.14	53	0.58	14	0.25	6	0.28	7	0.56	14	0.25	6	1.09	27	4.06
	Scenarios F	0.93	62	0.28	19	0.11	7	0.05	3	0.08	6	0.05	3	0.17	12	1.49

All estimates are based on a model climate sensitivity of ca. 2°C for a doubling of CO_2.
[1] Equilibrium temperature response calculations based on trace-gas concentration scenarios given by Wuebbles et al. (1984).
[2] Equilibrium temperature response calculations based on empirical expressions describing both observed and projected concentration changes as given by Wigley (1987).
[3] See Table 2.2.
[4] See Table 2.3.

The discrepancies likely reflect differences in the empirical expressions derived for observed and projected concentrations.

The projections over the next 50 years depend on the adopted scenarios. The estimates by Ramanathan et al. (1985) and Tricot and Berger (1987) are close to the medium and low scenarios C and D in our study. The "Efficiency" scenarios E and F for fossil fuel use cut the warming contribution from CO_2 to about 50 to 60 percent of the estimates obtained by the other authors.

Total Warming Commitment Including Ozone. To obtain the total equilibrium surface warming, we add to the figures shown in Tables 2.5 and 2.6 the greenhouse warming caused by ozone, as estimated by Ramanathan et al. (1985). This contribution amounts to 0.04°C from tropospheric O_3 for 1880–1980, and 0.14°C from both tropospheric and stratospheric O_3 for 1980–2030.

Based on the estimates of the various studies, we then obtain the following ranges for the total warming commitment (climate sensitivity about 2°C):

- 0.8–0.9°C for the period 1850/80–1980.
- 0.7–3.3°C for the period 1980–2030.
- 1.5–4.2°C for the total period 1850/80–2030.

An approximate threefold amplification toward the poles would yield a warming of

- 5–13°C for 1880–2030.

These ranges underline that similar findings for historic warming do not preclude widely differing results for future warming. There is not one, but several paths into the future. Which path will become reality is ultimately a function of policy decisions.

Comparison with Climate-Stabilization Targets. If we compare these warming impacts with the climate-stabilization targets discussed in Chapter 1, the following observations can be made:

- In the "low" combined scenarios, the warming commitment would increase by about 1.5°C in 2030 relative to preindustrial conditions. If realized, the climate would be about as warm as in the Holocene some 6000 years before present (yBP).

- If the warming commitment of "medium" combined scenarios were realized, the climate would warm by 2.2–2.6°C, that is, to levels not experienced since the last interglacial period (i.e., the Eem-Sangamon) some 125,000 yBP.

- In the "high" combined scenario A (no Montreal Protocol), the earth could warm to conditions not experienced since 3–5 million yBP (the Pliocene warm period), when it was some 4°C warmer than today.

3. Radiative Forcing Limit

Our equilibrium response calculations are based on a moderate-to-low climate sensitivity of 2.3°C. To comprehensively evaluate these scenarios against the warming limit, we now determine for the full range of climate sensitivities what equivalent CO_2 concentration would produce an equilibrium temperature rise equal to the warming limit of 2.5°C. These equivalent CO_2 concentrations can then be compared with the actual equivalent concentrations in each scenario (Table 2.4). We can then determine the maximum climate sensitivity under which each scenario would still satisfy the warming limit.[6]

Table 2.7 shows the maximum radiative forcings and equivalent-CO_2 concentrations for climate sensitivities ranging from 1.5 to 5.5°C, based on the formulation in Appendix 1 and linear scaling.

A comparison of this table with the equivalent CO_2 concentrations in Table 2.4 shows the following:

- Present equivalent CO_2 concentrations already preclude climate stabilization if climate sensitivity should be 5.5°C or more.
- The maximum climate sensitivity for which the lowest scenario (F) would still satisfy the climate stabilization criterion is 2.5 × 2.3/1.5 = 3.8°C (2 × CO_2).

Further emissions beyond 2030 would create additional commitments. But the above equilibrium warming figures do not tell us how fast these ultimate temperature levels would be reached. Thus, the rate of warming over time, which is an important parameter in climate stabilization, cannot be determined from these results. A full assessment of each scenario therefore requires calculation of the transient response.

---

[6]Assuming that concentrations remain level until equilibrium is reached.

**TABLE 2.7 Maximum Radiative Forcing and
Maximum Equivalent CO_2 Concentrations under
a Warming Limit of 2.5°C, as a Function of Model
Climate Sensitivity**

	Climate sensitivity (° C for $2 \times CO_2$)			
	1.5	2.3	4.5	5.5
Maximum Radiative forcing (W/m²)	7.32	4.77	2.44	2.00
Maximum equivalent CO_2 conc. (ppm)	911	610	422	393

See Appendix 2 for derivation.

4. Transient Response Studies

To estimate the delayed long-term warming effects, we now turn to a
transient response analysis. The time frame for this analysis spans the
historical period of 1860 to 1980 and the projected period of 1980 to
2100. The modeling concept used in the transient response experiments is shown in Figure 2.2 above.

Trace-Gas Concentrations to 2100

Carbon Dioxide. The estimated CO_2 concentration changes for the
extrapolated energy scenarios "Oak Ridge A" and "Energy Efficiency"
are shown in Table 2.1. The enormous fossil fuel consumption in the
"Oak Ridge A" scenario would lead to a more than sevenfold increase
in CO_2 concentration from the 1980 value of 338 ppm to almost 2500
ppm in 2100. In the extrapolated "IIASA Low" scenario (not shown in
Table 2.1), CO_2 concentration would double. In the "Energy Efficiency" scenario, the CO_2 concentration would peak at around 380
ppm in the late 2020s and then decline to 364 ppm by 2100, or about
8 percent above its level in 1980.

Other Trace Gases. We use a medium and a low scenario. In both the
medium and the low scenario, CFC-11 and CFC-12 emissions are cut
to 50 percent of 1986 production levels, equivalent to implementing
the Montreal Protocol without use of exemptions. Ongoing emissions

lead to a slow further rise in atmospheric concentrations. In 2100, the two gases contribute a CO_2 equivalent of 7 and 14 ppm, respectively.[7]

CH_4 and N_2O concentration scenarios are based on extrapolating case C in Table 2.4 for the medium case. In the low scenario, equivalent CO_2 concentrations for methane reach 31 ppm in 2100, and 8 ppm for nitrous oxide.

Combinations. We calculate three combined scenarios. In the combined high case, the "Oak Ridge A" scenario for CO_2 is combined with the medium scenario for other trace gases. In the combined medium case, the "IIASA Low" extrapolation for CO_2 is combined with the same medium trace-gas scenario. In the combined low case, the "Energy Efficiency" scenario is combined with the low assumptions for other trace gases. In this latter combined scenario, equivalent CO_2 concentrations peak around 2030 and then slowly decline to about 430 ppm by 2100. For reasons discussed below, the low scenario is also dubbed the "toleration" scenario.

Transient Warming Results. We calculate the transient warming ΔT_s for both a 1.5°C and a 4.5°C sensitivity. The results for two of the three combined scenarios are shown in Tables 2.8 and 2.9, respectively. All three combinations are graphically presented in Figures 2.10 and 2.11.

Across all scenarios, the key findings (relative to the 1850/60 mean global temperature) can be summarized as follows:

- In the more optimistic case (low climate sensitivity), average global transient warming in the year 2100 could range from 0.9 to 4.5°C.
- In the more pessimistic case (high climate sensitivity) the warming could range from 2.1 to 8.3°C.
- The total uncertainty in this warming trumpet reaches 7.4°C by 2100.
- The model-sensitivity-related uncertainty is 1.2°C (low scenario) to 3.8°C (high scenario).

[7]Equivalent CO_2 concentrations for individual trace gases are calculated on an incremental or marginal basis, assuming no other trace gases besides CO_2 are present. Mathematically, these ppm figures are not additive (see Appendixes 1 and 2). For the ranges of concentration values in the "low" scenario, the error from adding individual increments rather than applying the calculation to combinations of trace gases happens to be small.

TABLE 2.8 Transient Surface Temperature Change ΔT_e (K) for "Oak Ridge A" Energy Scenario and "Medium" Trace-Gas Scenarios and Different Climate Sensitivities T_e (K)

		"Oak Ridge A" energy scenario and "Medium" trace gas scenario							
T_e(K)	year	ΔT_e(K) (relative to 1860) due to					Total of all gases ΔT_e(K) relative to		
		CO_2	CH_4	N_2O	CFC-11	CFC-12	1860	1946–60	1980
	1860	0.00	0.00	0.00	0.00	0.00	0.00	-0.35	-0.51
	1980	0.31	0.12	0.02	0.02	0.04	0.51	0.16	0.00
	1990	0.38	0.15	0.04	0.02	0.04	0.63	0.28	0.12
	2000	0.52	0.16	0.05	0.02	0.04	0.79	0.44	0.28
	2010	0.68	0.17	0.06	0.02	0.04	0.98	0.63	0.47
	2020	0.87	0.18	0.07	0.03	0.04	1.20	0.86	0.69
1.5	2030	1.13	0.19	0.08	0.03	0.04	1.47	1.12	0.96
	2040	1.40	0.21	0.08	0.04	0.05	1.79	1.44	1.28
	2050	1.75	0.22	0.09	0.04	0.06	2.16	1.81	1.65
	2060	2.16	0.24	0.09	0.04	0.06	2.58	2.23	2.07
	2070	2.58	0.25	0.10	0.04	0.07	3.04	2.70	2.53
	2080	3.05	0.27	0.10	0.04	0.07	3.54	3.20	3.03
	2090	3.53	0.29	0.11	0.04	0.07	4.05	3.71	3.54
	2100	3.98	0.31	0.11	0.04	0.07	4.51	4.22	4.00
	1860	0.00	0.00	0.00	0.00	0.00	0.00	-0.64	-0.95
	1980	0.58	0.23	0.04	0.04	0.06	0.95	0.31	0.00
	1990	0.74	0.29	0.04	0.04	0.06	1.16	0.52	0.21
	2000	0.95	0.31	0.05	0.04	0.08	1.43	0.79	0.48
	2010	1.24	0.33	0.07	0.04	0.08	1.76	1.12	0.81
	2020	1.57	0.35	0.10	0.04	0.09	2.16	1.52	1.21
4.5	2030	2.03	0.37	0.12	0.04	0.09	2.64	2.00	1.69
	2040	2.51	0.40	0.13	0.05	0.10	3.20	2.56	2.25
	2050	3.09	0.44	0.15	0.06	0.11	3.86	3.22	2.91
	2060	3.80	0.47	0.17	0.06	0.12	4.61	3.97	3.66
	2070	4.56	0.51	0.18	0.07	0.13	5.45	4.81	4.50
	2080	5.43	0.54	0.20	0.07	0.13	6.37	5.73	5.42
	2090	6.34	0.57	0.21	0.07	0.14	7.33	6.69	6.38
	2100	7.26	0.61	0.22	0.08	0.15	8.33	7.69	7.38

Rounding errors with summation.

In terms of how much warming would come about in the worst case, the key conclusions are as follows:

- Even under a low (1.5°C) climate sensitivity, by 2100 the high combined scenario based on "Oak Ridge A" would warm the earth beyond all historic precedents over the last 5 million years (including the Pliocene, from 3 to 5 MyBP).
- The same conclusion applies to the medium combined scenario based on "IIASA Low," assuming a higher (4.5°C) climate sensitivity.
- Should the higher climate-model sensitivity of 4.5°C turn out to be correct, the realized warming by the year 2100 in the "Oak

TABLE 2.9 Transient Surface Temperature Change ΔT_s (K) for "Toleration" Scenario
("Energy Efficiency" Scenario and "Low" Trace-Gas Scenarios) and Different Climate Sensitivities T_e (K)

		\"Toleration\"-scenario (\"Efficiency\" energy scenario and \"Low\" trace gas scenario)							
T_e(K)	year	T_s(K) (relative to 1860) due to					Total of all gases ΔT_s(K) relative to		
		CO_2	CH_4	N_2O	CFC-11	CFC-12	1860	1946–60	1980
	1860	0.00	0.00	0.00	0.00	0.00	0.00	-0.35	-0.51
	1980	0.31	0.12	0.02	0.02	0.04	0.51	0.16	0.00
	1990	0.38	0.14	0.04	0.02	0.04	0.62	0.27	0.11
	2000	0.43	0.15	0.05	0.02	0.05	0.71	0.37	0.20
	2010	0.47	0.16	0.06	0.03	0.05	0.78	0.43	0.27
	2020	0.49	0.16	0.07	0.03	0.05	0.82	0.48	0.31
1.5	2030	0.51	0.17	0.07	0.03	0.06	0.85	0.51	0.34
	2040	0.52	0.17	0.07	0.04	0.06	0.87	0.52	0.36
	2050	0.52	0.17	0.08	0.04	0.06	0.88	0.53	0.37
	2060	0.51	0.18	0.08	0.04	0.07	0.89	0.54	0.38
	2070	0.50	0.18	0.08	0.04	0.08	0.89	0.55	0.38
	2080	0.50	0.18	0.09	0.04	0.08	0.90	0.55	3.39
	2090	0.50	0.18	0.09	0.04	0.08	0.90	0.55	0.39
	2100	0.50	0.18	0.09	0.04	0.08	0.90	0.55	0.39
	1860	0.00	0.00	0.00	0.00	0.00	0.00	-0.64	-0.95
	1980	0.58	0.23	0.04	0.04	0.06	0.95	0.31	0.00
	1990	0.73	0.27	0.05	0.04	0.06	1.15	0.51	0.20
	2000	0.87	0.30	0.06	0.04	0.07	1.34	0.70	0.39
	2010	0.98	0.31	0.08	0.04	0.07	1.49	0.85	0.54
	2020	1.06	0.33	0.09	0.05	0.08	1.62	0.98	0.67
4.5	2030	1.11	0.35	0.10	0.06	0.09	1.72	1.08	0.77
	2040	1.14	0.37	0.12	0.06	0.10	1.79	1.15	0.84
	2050	1.16	0.38	0.14	0.06	0.11	1.86	1.22	0.91
	2060	1.18	0.39	0.15	0.07	0.12	1.91	1.27	0.96
	2070	1.18	0.41	0.16	0.07	0.13	1.95	1.31	1.00
	2080	1.18	0.42	0.17	0.08	0.13	1.98	1.34	1.03
	2090	1.18	0.43	0.18	0.08	0.14	2.02	1.38	1.07
	2100	1.18	0.44	0.19	0.08	0.15	2.05	1.41	1.10

Rounding errors with summation.

Ridge A" case would even go beyond that of the Oligocene/Miocene period, some 38 to 15 MyBP.

- Since a major portion of the equilibrium warming is disguised by the inertia of the ocean, the ultimate temperature rise in the years beyond 2100 would be even higher.
- The upper limit for the global average warming rate of 0.1°C per decade, as set out in Chapter 1, will be exceeded four to six times in the high scenario.

Focusing on the optimistic end of the range of outcomes, the following observations apply:

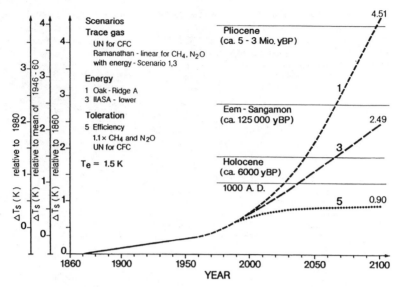

Figure 2.10   *Transient surface-temperature change for various scenarios related to different time periods and paleoclimatic conditions.*

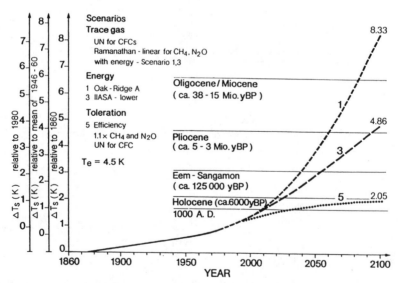

Figure 2.11   *Transient surface-temperature change for various scenarios related to different time periods and paleoclimatic conditions.*

- In the low case based on the "Efficiency" scenario, with rapid CO_2 control and significant reductions in the CH_4, N_2O, and CFC growth rates, the realized global average warming by 2100 would range between about 1 and 2°C;
- If the lower climate sensitivity should prove correct, the low combined scenario would not reach the conditions of the Holocene period some 6000 years ago.
- Assuming the higher climate sensitivity is correct, the realized warming would reach the range of conditions last experienced in the Holocene period. By 2100, realized warming would reach the conditions of the last interglacial period, the Eem-Sangamon, some 125,000 yBP.
- In the low-sensitivity/low scenario case, the average warming rate over the next 120 years would be lower than the warming rate during the last 130 years. The trend curve of transient warming would be fully stabilized during the second half of the next century.
- In the high-sensitivity/low scenario case, the target warming rate of 0.1°C per decade would be exceeded by about 20 percent. Transient temperatures would still be rising slightly in 2100, albeit at much less than the target rate (Table 2.9).
- The low combined scenario would roughly stabilize the trend curve of realized warming around 2100, in contrast to the unbroken upward trends for the other scenarios.

The latter point returns us to the relationship between warming commitment and realized warming, which was discussed in Section C.

Comparison of Transient with Equilibrium Temperature Change. It is instructive to calculate for the scenarios and modeling approach used in this study how much of the warming commitment made in a particular year is realized in that year. Table 2.10 shows a comparison of the expected equilibrium warming T_e with the transient warming T_s for the high and low scenarios. The ratio of T_s/T_e gives the realized-warming fraction, that is, the amount of global mean surface warming that has already occurred. To be able to observe trends, the figures are given for a series of progressively longer time periods from 1850/60 to 2030.[8]

---

[8]Equilibrium temperatures were calculated using the methods in Appendixes 1 and 2, including linear scaling among model climate sensitivities.

TABLE 2.10   Ratio of Transient and Equilibrium Warming for Climate Sensitivity 4.5°C[1, 2]

Period	High Scenario			Low Scenario		
	T_e	T_s	T_s/T_e (4.5)	T_e	T_s	T_s/T_e (4.5)
1850–1980	1.71	0.95	0.55	1.71	0.95	0.55
1850–1990	2.25	1.16	0.52	2.17	1.15	0.53
1850–2000	2.83	1.43	0.51	2.57	1.34	0.52
1850–2010	3.57	1.76	0.49	2.65	1.49	0.56
1850–2020	4.46	2.16	0.48	2.71	1.62	0.60
1850–2030	5.48	2.64	0.48	2.72	1.72	0.63

[1]Equilibrium warmings calculated as shown in Appendix 2, linearly scaled to match transient model climate sensitivities.
[2]Transient warmings for higher scenario from Table 2.8, for lower scenario, from Table 2.9.

The modeling calculations, which are normalized to a climate sensitivity of $4.5°CO_2$, suggest that for the historic period from 1850 to 1980, some 55 percent of the expected equilibrium warming should have already occurred. The range for the high and low scenarios for the period up to 2030 is 0.48–0.62. These revealed warming fractions are similar to those reported in other analyses.[9]

Emission trends also have some influence on the realized-warming fraction: with higher and growing trace-gas emissions, the share of the total warming commitment that remains hidden increases relative to scenarios of lower and declining emissions. These results underline that a strategy of early and major trace-gas reductions is important to minimize the legacy of unrealized warming commitments that will be borne by future generations.

Comparison with Other Transient Response Studies. Recently, Hansen et al. (1988) of the Goddard Institute for Space Studies performed a scenario analysis similar to the one described in this text, but using a more sophisticated modeling approach based on a 3-d general circulation model (GCM). The model had a climate sensitivity of 4.2°C. Transient calculations were performed by coupling the atmospheric GCM to an ocean model based on replicating, for each grid cell of the 3-d model, the simple diffusion approach used in this study.

The emission scenarios are broadly comparable to the ones used in this book. In the low scenario, CO_2 concentrations stabilize at 368 ppm, and CFC-11 and CFC-12 are phased out by 2000. CH_4 concen-

[9]Ramanathan (1988) reports a range of 45–75 percent for the realized-warming fraction from GCM-based calculations. See also Figure 2.3 and Section C.

trations stabilize at about 1900 ppm, and no increase occurs in N_2O abundance. The low scenario achieves constant climate forcing after the year 2000, somewhat earlier than in our calculations.

With these assumptions, Hansen et al. arrive at a transient warming of about 1.3–1.5°C relative to 1880 (0.5–0.7°C relative to the 1951–1980 mean) by about 2010–2030, with roughly stabilized temperatures thereafter. If we take into account the somewhat lower climate sensitivity and radiative forcing scenario used in their study, these more sophisticated 3-d model simulations thus arrive at very similar figures as the 4.5°C sensitivity runs based on the 1-d model used in this study.

5. Discussion

Our calculations above clearly illustrate the importance of the assumed climate sensitivity in translating a global warming limit into specific emission reduction requirements. As discussed in Chapter 1, and as suggested by the trend toward higher climate sensitivities in the more recent scientific work, a risk-minimizing climate policy should be based on the upper end of the climate-sensitivity range. Accordingly, we discuss and interpret our climate modeling results with focus on climate sensitivities in the 4.0–5.5°C range.

Our transient response calculations suggest that at least for the period until 2100, global surface warming could still be limited to about 2°C or less—significantly less than the 2.5° relative to 1850—provided that climate sensitivities fall within the range from 1.5 to 4.5°C. And because transient temperature is less than proportional to climate sensitivity (see Section C), the same conclusion can be extended to the entire range of current model climate sensitivities up to 5.5°C.

At the same time, equivalent CO_2 concentrations in the low scenario (Table 2.11) climb significantly beyond the allowable equilibrium limit of 393–423 ppm calculated for climate sensitivities of 4.5–5.5°C (see Table 2.7), and decline only slowly by a moderate 5 percent after reaching their peak around 2030. In 2100, the equivalent CO_2 concentration is still about 430 ppm. Therefore, for climate sensitivities in the neighborhood of 4.5–5.5°C, the 2.5°C warming limit would eventually be reached and exceeded should this concentration level persist. If, on the other hand, equivalent CO_2 concentrations were to decline further toward present levels during the first few decades of the twenty-second century, the realization of more than a 2.5°C warming could possibly be avoided. With this assumption, and for a climate sensitivity in the neighborhood of 4.0–4.5°C, the permissible peak

TABLE 2.11 Equivalent CO_2 Concentrations in the "Toleration" Scenario

Gas	Equivalent CO_2 Concentration [1]	
	ca. 1990	ca. 2100
CO_2	355	364
CH_4	28	31
N_2O	6	8
CFC-11	3	7
CFC-12	6	14
Total	400	429

[1]Equivalent CO_2 concentrations calculated from Appendixes 1 and 2. Individual values for CH_4, N_2O, and the CFCs are calculated incrementally and therefore are not additive.

concentration limit during the twenty-first century might be as much as 450 ppm.

It is worth stating what such an eventual return of equivalent CO_2 concentrations to present levels would imply: fossil carbon dioxide emissions would have to be fully eliminated; a full ban on CFCs would have to be maintained without emitting new kinds of industrial greenhouse gases; the carbon storage in the land biota would have to be increased above current levels; and/or methane concentrations would have to be reduced somewhat rather than just stabilized. We discuss in Chapter 3 how these requirements square with our current knowledge of emission control options for the nonfossil trace gases.[10]

We can assess our warming results in yet another way. If the world should succeed in fully eliminating CFC emissions around the turn of the century, equivalent CO_2 concentrations would be reduced by about 10–15 ppm relative to the low scenario (see Table 2.11). On the other hand, the phaseout[11] assumptions for carbon dioxide by Lovins

[10]In allowing for a transient excursion of the total radiative forcing equivalent to concentration values significantly above 400 ppm, one should also bear in mind that the climate system might not move about between alternate states in the smooth and gradual fashion current climate models predict (Broecker 1987).

[11]In this text, we occasionally refer to reductions in fossil CO_2 emissions as a fossil phaseout. Note that in using this term in subsequent contexts we do not necessarily mean a complete elimination of these emissions. While such an assumption was made in the outer limit case of the "low" scenario, reductions to a low but significant residual emission level would be sufficient to eliminate fossil fuels as a driving force in climate change.

et al. (1981) were based on rapid policy action beginning in the early 1980s. They are probably overly ambitious from today's perspective, since such early action has not occurred. If we assume that CO_2 concentrations stabilize at the peak level of about 380 ppm rather than declining again, this would roughly cancel the increment from the early phaseout of CFCs. All in all, trace gases other than CO_2 would contribute an increment of about 50 ppm, and equivalent CO_2 concentrations would again be about 430 ppm.

6. Conclusions

These explorations lead to the following major conclusions:

- If actual climate response falls within the range of current model climate sensitivities, the goal of climate stabilization as defined in Chapter 1 could still be realized.
- The maximum allowable equivalent CO_2 concentrations in the next century would be 430–450 ppm, provided that concentrations decline again thereafter.
- The range of emission scenarios under which this goal could be achieved is very narrow. It is roughly captured by trace-gas emissions as assumed in the low scenarios, and requires major and rapid reductions in greenhouse gas emissions.
- Even the low scenario brings with it significant climatic risks, which are, however, probably unavoidable and must be tolerated.
- Middle-of-the-road scenarios that project moderate and slow changes in emission trends, including a transition to constant CO_2 emissions at present or higher levels, will not meet stabilization requirements. They would rapidly lead the world into climatic conditions not experienced since 3–5 MyBP;
- CO_2 remains the dominant trace gas in all scenarios. Mitigation of these releases must therefore be the backbone of any climate-stabilization policy.

Because the range of warming expected from the low scenario approximates climate stabilization at minimum avoidable, but nonzero climatic risk, we have dubbed it the "toleration" scenario. The term does not only apply to the specific combination of emissions described above, but also to any other scenario that is similarly compatible with the warming limit.

E. FROM CONCENTRATION CEILINGS TO POLICY TARGETS

The ominous implications of these and other recent climate-modeling results have led to calls for deliberate reductions in greenhouse gas emissions. Before delving into the details of how such reductions could be brought about, it is useful to take a broad look at the types of measures that can be taken, and the policy fields relevant to the greenhouse effect.

1. Abatement Approaches and Policy Fields Involved in Climate Stabilization

The principal "technical fixes" that can be used to reduce trace-gas emissions include CFC control, recycling, and substitution; more efficient energy use; substitution of fossil fuels by nonfossil energy sources; substitution of fossil fuels by other fossil fuels; emission reduction at the source; tree planting and, more generally, the control of deforestation and soil destruction; and the modification of agricultural production patterns. Table 2.12 shows the trace gases for which each approach is relevant and applicable.

Implementing these measures involves a broad set of policy-making areas, such as the licensing and regulation of industrial chemicals; regulations governing stationary and mobile air pollution emissions in industry, households, and transport activities; energy planning, utility regulation, and fuel taxation; regulation of imports of tropical forest products; agricultural subsidies and farm support policies; land-use zoning and planning; land reform, and development assistance policies.

For most airborne pollutants, the dominant approach to emission control so far has been emission reductions at the source. However,

TABLE 2.12   Applicability of Control Measures for Trace Gases

Control option	GHG that is reduced or influenced
Control of industrial chemicals	CFC-11, CFC-12, other CFCs, Bromine compounds, O_3S
More efficient energy use	CO_2, N_2O, CO, NOx, HC, OH, CH_4, O_3T, S
Substitution with non-fossil sources	same
Interfossil fuel substitution	mainly CO_2, NOx, OH, O_3T
Emission reduction at the source	NO_x, CO, HC, OH, O_3T
Control deforestation/soil destruction	CO_2, N_2O, CO, NO_x, HC, OH, CH_4, O_3T, S
Modification of agric. prod. methods	CO_2, N_2O, CO, NO_x, OH, CH_4, O_3T, S
O_3T = ozone in troposphere; O_3S = ozone in stratosphere.	

inspection of Table 2.12 shows that the most important trace gases, such as carbon dioxide, cannot be controlled in that manner. Stack gas scrubbing could possibly be applied to some large stationary emitters such as power plants, but the costs of doing so are prohibitive, and disposal problems unsolved. This explains why energy policy must be the backbone of climate stabilization.

2. Development of Emission Reduction Targets

Because of the dominance of fossil fuel consumption in global climate warming, we focus on the development of global and regional control targets for fossil carbon dioxide. The toleration scenario defines the gross proportions climate-stabilizing energy strategies will need to satisfy. But many important questions remain about the exact speed of fossil fuel displacement that is needed or feasible in the near term and over the long term, how near-term choices could preempt the attainment of fossil carbon budgets over the long term, the relative contributions to emission reductions of industrialized and developing countries, and the best mix of fossil-substituting resources. These issues are explored in Chapter 4 and subsequent chapters.

Compared to the sources for most other trace gases, the magnitude of fossil fuel consumption and the technologies and applications involved are relatively well understood. Alternative phaseout schedules along the lines of Figure 1.3 can therefore be explored in terms of their practical implications. But to make this detailed discussion of a fossil CO_2 phaseout climatically meaningful, we also must define the major practical steps needed to limit radiative forcing from nonfossil trace gases to the approximate magnitudes indicated in the toleration scenario.

In Section E of Chapter 1, we outlined the warming limit/emission budget/reduction schedule methodology (WERM). At this point we have specified the total radiative forcing and equivalent CO_2 concentrations that could be compatible with the warming limit (step 5), and we have established initial orders of magnitude for the equivalent CO_2 contribution from trace gases other than carbon dioxide.

In our further discussion, a principal objective is to establish the degree of optimism involved when one assumes that warming from trace gases other than carbon dioxide can be controlled along the lines of the "toleration" scenario. This is an important assessment for the interpretation of all subsequent energy policy discussions: the more formidable the control requirements, the greater the risk that they will not be fully achieved in practice. This, in turn, will put more stringent demands on energy policy if one and the same warming limit is to be observed. Our entire study is, therefore, conceptualized to define an upper limit for future fossil fuel consumption, and corresponding minimum requirements for energy policy responses.

Our discussion so far shows that the purely mathematical options for trade-offs that apparently exist among warming contributions from each trace gas are more limited and less relevant when practical constraints are taken into account. For one, the toleration scenario requires close to the maximum potential of emission reductions in each trace-gas area. Secondly, some sources, such as CFCs, are more easily or reliably controlled than others, such as methane or nitrous oxide. Finally, measures and policies for each trace gas must be in harmony with the overall goal for sustainable world development.

Without such assurance, climate-stabilization policies run the risk of becoming yet another linear single-issue effort in an interlinked world. It is hence important to identify and pursue those measures that simultaneously address problems other than climate change. Our analysis in Chapter 3 seeks to identify the most important opportunities for such tie-ins (Schneider 1989).

REFERENCES

Bach, W. (1985), "CO$_2$-Zunahme und Klima. Modeller-gebnisse," *Geoökodynamik*, Vol. 6, No. 3, pp. 229–292.

Bach, W., and H. J. Jung (1987), "Untersuchung der Beeinflussung des Klimas durch Spurengase mit Hilfe von Modellrechnungen," *Munstersche Geographische Arbeiten*, Vol. 26, pp. 45–64.

Bach, W., et al. (1985), "Modeling the Influence of Carbon Dioxide on the Global and Regional Climate. Methodology and Results," *Münstersche Geographische Arbeiten*, No. 21.

Bolin, B., R. Doos, J. Jaeger, and R. A. Warrick, eds. (1986), *The Greenhouse Effect: Climatic Change and Ecosystems*, Wiley, New York.

Broecker, W. S. (1987), "Testimony before the U.S. Senate Subcommittee on Environmental Protection," Jan. 28, Washington, D.C.; see also *Nature*, Vol. 328 (1987), p. 123.

CDAC (Carbon Dioxide Assessment Committee) (1983), *Changing Climate*, National Research Council, National Academy Press, Washington, D.C.

Cess, R. D., and S. D. Goldenberg (1981), "The Effect of Ocean Heat Capacity upon Global Warming due to Increasing Atmospheric Carbon Dioxide," *Journal of Geophysical Research*, Vol. 86, pp. 498–502.

Dickinson, R. E. (1986), in B. Bolin, R. Doos, J. Jaeger, and R. A. Warrick, eds., *The Greenhouse Effect: Climatic Change and Ecosystems*, Wiley, New York, pp. 207–270.

Dickinson, R. E., and R. Cicerone (1986), "The Climate System and Modeling of Future Climate," in B. Bolin, R. Doos, J. Jaeger, and R. A. Warrick, eds. *The Greenhouse Effect: Climatic Change and Ecosystems*, Wiley, New York, pp. 206–270.

Edmonds, J. A., J. Reilly, J. R. Trabalka, and D. E. Reichle (1984), *An Analysis of Possible Future Atmosphere Retention of Fossil Fuel CO_2*, DOE/TR/013, U.S. Department of Energy, Washington, D.C.

Emanuel, W. R., et al. (1984), *Computer Implementation of a Globally Averaged Model of the World Carbon Cycle*, DOE/TR/010, U.S. Department of Energy, Washington, D.C.

Fabian, P. (1986), "Halogenated Hydrocarbons in the Atmosphere," in O. Hutzinger, ed., *The Handbook of Environmental Chemistry*, Vol. 4, Part A: Air Pollution, Springer, Berlin, pp. 23–51.

Gammon, R. H., E. T. Sundquist, and P. J. Fraser (1985), "History of Carbon Dioxide in the Atmosphere," in J. R. Trabalka, ed., *Atmospheric Carbon Dioxide and the Global Carbon Cycle*, U.S. Department of Energy, Washington, D.C., pp. 25–62.

Haefele, W., et al. (1981), *Energy in a Finite World*, Cambridge, MA.

Hammitt, J. K., et al. (1986), *Product Uses and Market Trends for Potential Ozone-Depleting Substances: 1985–2000*, The Rand Corporation, Santa Monica, CA.

Hansen, J. E. (1983), in S. Seidel and D. Keyes, *Can We Delay a Greenhouse Warming?* U.S. Environmental Protection Agency, Washington, D.C.

Hansen, J. E. (1989), Presentation at the American Geophysical Union Fall 1989 Meeting, San Francisco, Dec.

Hansen, J., et al. (1981), "Climatic Impact of Increasing Atmospheric Carbon Dioxide," *Science*, Vol. 213, pp. 957–966.

Hansen, J. E., et al. (1985), "Climate Response Times: Dependence on Climate Sensitivity and Ocean Mixing," *Science*, Vol. 229, pp. 857–859.

Hansen, J. E., et al. (1988), "Global Climate Changes as Forecast by the Goddard Institute for Space Studies Three-Dimensional Model," *Journal of Geophysical Research*, Vol. 93, pp. 9341–9364.

Hoffman, J., and M. J. Gibbs (1988), *Future Concentrations of Stratospheric Chlorine and Bromine*, EPA 400/1-88/005, U. S. Environmental Protection Agency, Office of Air and Radiation, Washington, D.C.

Jones, P. D., et al. (1986), "Global Temperature Variations between 1861 and 1984," *Nature*, Vol. 322, pp. 430–434.

Keepin, B., and B. Wynne (1984), "Technical Analysis of IIASA Energy Scenarios," *Nature*, Vol. 312, pp. 691–695.

Khalil, M. A. K., and R. A. Rasmussen (1981), "Increase in the Atmospheric Concentrations of Halocarbons and N₂O," in J. J. Delᵢisi, ed., *Geophysical Monitoring for Climatic Change*, No. 9, U.S. Department of Commerce, Washington, D.C., pp. 134–139.

Lashof, D. A. (1989), "The Dynamic Greenhouse: Feedback Processes That May Influence Future Concentrations of Atmospheric Trace Gases and Climate Change," *Climatic Change* (forthcoming); Revised draft, Jan. 1989.

Lashof, D. A., and D. A. Tirpak, eds. (1989), *Policy Options for Stabilizing Global Climate*, Draft report to the U.S. Congress, U.S. Environmental Protection Agency, Office of Policy, Planning, and Evaluation, Feb.

Lovins, A. B., L. H. Lovins, F. Krause, and W. Bach (1981), Least-Cost Energy: Solving the CO₂-Problem, Brick House, Andover, MA. (German version: *Wirtschaftlichster Energieeinsatz: Lösung des CO₂-Problems*, Karlsruhe, FRG, 1983.)

Mintzer, I.M. (1987), *A Matter of Degrees: The Potential for Controlling the Greenhouse Effect*, Research Report No. 5, World Resources Institute, Washington, D.C.

Oeschger, H., U. Siegenthaler, U. Schatterer, and A. Gugelmann (1975), "A Box Diffusion Model to Study the Carbon Dioxide Exchange in Nature," *Tellus*, Vol. 27, pp. 168–192.

Ramanathan, V. 1988: "The Greenhouse Theory of Climate Change: A Test by an Inadvertant Global Experiment," *Science*, Vol. 240, (Apr. 15), pp. 293–299.

Ramanathan, V., et al. (1985), "Trace Gas Trends and Their Potential Role in Climate Change," *Journal of Geophysical Research*, Vol. 90, pp. 5547–5566.

Ramanathan, V., et al. (1987), "Climate-Chemical Interactions and Effects of Changing Atmospheric Trace Gases," *Reviews of Geophysics*, Vol. 25, No. 7, pp. 1441–1482.

Rasmussen, R. A., and M. A. K. Khalil (1986), "Atmospheric Trace Gases: Trends and Distributions over the Last Decade," *Science*, Vol. 232, pp. 1623–1624.

Schneider, S. H. (1989), "The Greenhouse Effect: Science and Policy," *Science*, Vol. 243, No. 4892, (Feb. 10), pp. 771–781.

Tricot, C., and A. Berger (1987), "Modeling the Equilibrium and Transient Response of Global Temperature to Past and Future Trace Gas Concentrations," *Climate Dynamics*, Vol. 2, pp. 39–61.

Weiss, R. F. (1981), "The Temporal and Spatial Distribution of Tropospheric Nitrous Oxide," *Journal of Geophysical Research*, Vol. 86, pp. 7185–7195.

Wigley, T. M. L. (1984), "Carbon Dioxide, Trace Gases and Global Warming," *Climate Monitor*, Vol. 13, No. 5, pp. 133–148.

Wigley, T. M. L. (1987), "Relative Contributions of Different Trace Gases to the Greenhouse Effect," *Climate Monitor*, Vol. 6, No. 1, pp. 14–28.

WMO (1985), *Atmospheric Ozone 1985*, 3 Vols., Global Ozone Research and Monitoring Project, Report No. 16, World Meterological Organization, Geneva.

Wuebbles, D. J., M. C. MacCracken, and F. M. Luther (1984), *A Proposed Reference Set of Scenarios for Radiatively Active Atmospheric Constituents*, Technical Report .015, U.S. Department of Energy, Washington, D.C.

Ziegler, A., and R. Holighaus (1979), "Gas, Kraftstoff und Heizöl aus Kohle," *Umschau*, Vol. 79, No. 12, pp. 367–376.

3

Control Requirements
for Nonfossil
Greenhouse Gases

A. INTRODUCTION

This chapter deals with trace gases other than CO_2 from fossil fuel burning. We investigate what technical and policy measures could be employed to realize the concentration targets of the "toleration scenario"; how difficult it might be to realize these scenario targets; and what trade-offs, if any, could be made among greenhouse gases for which reductions are easier and those for which they are more difficult and uncertain. We include in this chapter fossil carbon dioxide emissions from cement production, and carbon monoxide and methane releases from fossil energy production and use.

In the various subsections below, a brief overview is provided of the global chemical cycle for each trace gas. A summary for the five major radiatively active gases is given in Table 3.1. The table shows their warming contributions as calculated in the toleration scenario, their major sources, and current emission rates. Also shown are several columns with percentage reductions. The first of these indicates the approximate reduction required to stabilize atmospheric concentrations as estimated by Lashof and Tirpak (1989). The next column shows the range of reductions that could be feasible for various emission sources and trace gas uses. The corresponding control options are the subject of this chapter. The final column shows the reductions relative to present emission levels as assumed in the "low" scenario of Chapter 2.

TABLE 3.1 Estimates of Trace-Gas Emission Reduction Potentials by Individual Source or Use, and Assumed Reduction in the "Toleration" Scenario

Gas	Anthropogenic source/use	Emission rate base year [1]	Reduction 2100 vs. base year [1] To reach stable concent.[2]	Estimated technical potential	Toleration scenario	ΔT_s(K) [3] for T_s=4.5K in 2100
		Mt/yr [5]	%	%	%	K
CO_2	Cement [4]	130				
	Coal [4]	2173				
	Oil [4]	2182				
	Gas [4]	853				
	Fossil fuel use [4]	5338				
	Land use conversion	1000–2600				
	Total	6300–7900	50–80	100	100	1.18
		Mt/yr	%	%	%	K
CH_4	Cattle	65–100		20		
	Biomass burning	20–80		20–60		
	Rice paddy fields	25–170		20		
	Landfills	20–70		50–70		
	Natural gas & mining losses	40–100		100		
	Total	170–520	10–20	50–60	10–20	0.44
		Mt/yr	%	%	%	K
N_2O	Fossil fuel combustion	0.1–0.3		100		
	Fertilization/Cultivation of soils	0.01–2.2		0–75		
	Biomass burning	0.2		50–75		
	Total	4.6–10.5	80–85	40–75	60–70 [6]	0.19
		kt/yr [5]	%	%	%	K
CFC-11	Foam	181.5			80	
	Aerosol	102.3			95	
	Refrigeration & air condition	26.4			50	
	Other uses	19.8			58	
	Total	330.0	75–100	100	81	0.08
		kt/yr	%	%	%	K
CFC-12	Refrigeration & air condition	215.6			75	
	Aerosol	140.8			95	
	Foam	57.2			75	
	Other uses	26.4			48	
	Total	440.0	75–100	100	80	0.15
TOTAL						2.05

[1] Estimates with error bands refer to any year between 1970 and 1990; the fossil CO_2 and CFC estimates refer to 1985; data are from Table 3.9 and Marland et al. (1988) and Houghton et al. (1987) (CO_2), Table 3.6 (CH_4), Table 3.4 (N_2O), and Table 3.2 (CFC-11 and CFC-12).
[2] Estimates from Lashof and Tirpak (1989).
[3] ΔT_s values calculated for the "toleration" scenario (i.e., "Efficiency" scenario for energy and low trace-gas scenario).
[4] Data are from Marland et al. (1988). CO_2 is given as C.
[5] Mt = 10^6 tons; kt = 10^3 tons.
[6] Calculations based on Weiss (1981), assuming a lifetime of 170 years.

To better mesh policy discussions with the list of gases to be addressed, we first investigate all those emissions that are mainly associated with the urban activities of modern industrial societies (Section B).

Next, those sources are examined that are mainly related to agricultural production methods and other rural activities (Section C). Here, the focus is on altered agricultural production methods and food consumption patterns.

Section D explores the contribution of biospheric carbon dioxide to the global carbon cycle and to rising atmospheric carbon dioxide concentrations. In Section F, those forestry measures and programs are required to reduce climate risks are investigated.

In all of our discussions, we pay special attention to issues of sustainable development and the needs of Third World countries. Sections C, D, and E as a unit address the key issue of increasing food production while preserving and restoring biospheric carbon pools.

B. CFC RELEASES AND OTHER URBAN – INDUSTRIAL EMISSIONS

The principal radiatively active gases in this category are the chlorofluorocarbons CFC-11 and CFC-12, nitrous oxide (N_2O), and methane (CH_4). Table 3.1 shows their contribution to global transient warming in the year 2100 under the toleration scenario. Collectively, the urban–industrial sources represent about 20 percent of the total 2100 transient warming.

We also briefly consider sources and control options for the chemically active nitrogen oxides (NO_x) and carbon monoxide (CO).

1. Chlorofluorocarbons (CFCs)

In recent years, the average growth rate of chlorofluorocarbon emissions has been about 5 percent per year. Due to the complex chemistry and long residence time of several of the CFCs in the atmosphere, major (up to 100 percent) reductions in current release rates are required to stabilize atmospheric concentrations (Table 31). The key initiatives for realizing such reductions will have to come from the industrialized countries (ICs). These countries currently account for about 95 percent of global releases.

Reducing CFC emissions will both prevent climate warming and avoid the impacts that ozone depletion would have on health (especially skin cancer), on agricultural productivity, and on the phytomass supporting marine life in the oceans. As further discussed below, the need to control ozone depletion may impose even more stringent control requirements on these substances than the need to limit climate warming.

Sources of CFCs. Table 3.2 shows current sources and applications for the two most important substances, CFC-11 and CFC-12. Aerosol propellants, air conditioning and refrigerator systems, insulation and packaging foams, and industrial and dry cleaning solvents are the dominant uses. In Table 3.3, we list for these and ten other CFCs a number of important characteristics, including recent growth rates, natural decay periods, and relative greenhouse warming, and ozone depletion potentials. Also listed are three CFCs (CFC-22, CFC-132b, CFC-134a) that are not yet in wide use, but are being considered as substitutes for the most damaging commercial chemicals. Eight of the 12 chemicals shown are covered by the Montreal Protocol (see below). The exceptions are methyl chloroform, carbon tetrachloride, and the three substitute CFCs already mentioned.

There are gaps in the table reflecting the lack of current knowledge about some of these gases. Moreover, there are significant differences in the values of the relative greenhouse warming potential of individual gases given by Ramanathan et al. (1985) and those given by the Chemical Manufacturers Association (CMA 1987b).

Mitigation Options. The following control measures have been suggested (Miller and Mintzer 1986; EPA 1988):

Increase in Efficiency of CFC Use. For example, replacing rotary compressors by reciprocating compressors in refrigerators and chillers

TABLE 3.2   Global Sources for CFC-11 and CFC-12

Sources/uses	CFC-11 10^3t/yr	%	Sources/uses	CFC-12 10^3t/yr	%
Natural sources:	None		Natural sources:	None	
Man-made sources/uses:			Man-made sources/uses:		
Closed-cell foams	118.8	36	Refrig./air cond.	215.6	49
Aerosol propellant	102.3	31	Aerosol propellant	140.8	32
Open-cell foams	62.7	19	Closed-cell foams	35.2	8
Refrig./air cond.	26.4	8	Open-cell foams	22.0	5
Other uses	19.8	6	Other uses	26.4	6
Total	330.0	100		440.0	100

1 The estimates refer to 1985
From: WMO (1985); Hammitt et al. (1986); Wuebbles and Edmonds (1987)

TABLE 3.3 Summary of the Main CFC Characteristics Influencing the Ozone Layer and the Climate

Compound [1)2)]	Potential use [1)2)]	Estimated global production (emission) in 1985 [2)7)] (kt)	Lifetime yrs. [4)5)]	Concentration in troposphere in 1980 [5)] relative to		Growth in trop. % /yr [6)7)12)]	Relative greenhouse effect relative to CFC-12	Relative ozone depletion potential relative to CFC-1 [4)]	Share of the tot. contrib. to ozone depletion (%) [3)]
				ppb	CFC-12				
CFC-11 ($CFCl_3$)	Foam, aerosol, refrig.&air cond.	340 / −238	65	0.18	0.64	4–5	0.87[5)]	1	25.8
CFC-12 (CF_2Cl_2)	Refrig.,&air cond., aerosol, foam	440 / −412	130	0.28	1	5–6	1.0[6)]	0.9–1.1	44.7
CFC-113 ($C_2F_3Cl_3$)	Solvent, refrigerant	163.2 / −138	90	0.025	0.09	12–15	0.3–0.8[6)]	0.8	11.7
CFC-114 ($C_2F_4Cl_2$)	Foam, refrigerant	low	180	0.015	0.05	6		0.6–1.1	
CFC-115 (C_2F_5Cl)	Mixed with CFC-22 to form CFC-502 used as refrigerant	low	380	0.005	0.02		1.2[9)]	0.3–0.6	
CFC-22 (CHF_2Cl)	Refrigerant sterilant	210 / −72	20	0.06	0.21	8–11	0.31	0.05–0.08	0.4
CFC-132 b ($CH_2F_2CCl_2$)	Replacement for CFC-113						0.02[8)]	< 0.05[10)]	
CFC-134 a (CH_2CF_4)	Replacement for CFC-12						0.07[8)]	0[10)]	
Methyl Chloroform (CH_3CCl_3)	Solvent	544.6 / −474	0.7	0.14	0.5	5	0.13	0.01[8)]	5.1
Carbon Tetrachloride (CCl_4)	Used in CFC production and grain fumigation	1029 / −66	50	0.13	0.46	1.5	0.48	0.9–1.1	7.6
Halon 1301 ($CBrF_3$)	Fire extinguishant (permanent)	10.8 / −3	110	0.001	< 0.01		0.8[8)]	8–11	3.7
Halon 1211 (CF_2BrCl)	Fire extinguishant (portable)	10.8 / −3	25 (12–15)[11)]	0.001[11)]		20		2–3	0.9

1)Miller and Mintzer (1986). 2)Hammitt et al. (1986). 3)Hammitt et al. (1987). 4)CMA (1987a). 5)Ramanathan et al. (1985). 6) Enhalt (1988). 7)Fabian (1988). 8)CMA (1987b). 9)Reijnders (1987). 10)UNEP (1987). 11)Wuebbles and Edmonds (1987). 12)Rowland (1988).

could reduce the mass of refrigerant (especially CFC-12) by 50–70 percent or even more.

Reduction of Operating Losses. For instance, of the CFC-12 used for automobile air conditioning, typically 50 percent escapes during servicing, 30 percent is lost in routine leakage, and 20 percent escapes when cars are scrapped. All of this can be reduced.

Recovery and Recycling. About 50 percent of the CFC-11 used in foam manufacturing can be recovered through carbon filtration. CFC-12 used in vehicular air conditioning can also be recovered. CFC-113 used for degreasing and cleansing can be recycled through closed-cycle in-house distillation equipment.

Substitution by CFCs with Little or No Cl. A substitution of one CFC by another must be carefully considered by examining all factors involved, such as toxicity, flammability, and economics, and so forth, but above all the relative ozone-depletion potential (ODP), the contribution to the greenhouse effect, and other effects on the environment. For example, CFC-22 ($CHClF_2$) is considered a candidate for replacing CFC-12 in home and vehicular air conditioning. It has only about one-twentieth the ODP and one-fourteenth the greenhouse potential of CFC-12 (see Table 3.3).[1] But since it is largely broken down in the troposphere, the side effects of the decomposition products must be carefully examined prior to a shift. One of the decay products in the troposphere is hydrochloric acid (HCl), which is itself an ozone-depleting substance.

Substitution of CFC-11 and CFC-12 with Other Substances. CFCs can be replaced as aerosols and refrigerants by substances without any Cl, such as CFC-152a (CH_3CHF_2).

Switching to CFC-Free Processes and Products. For instance, the use of aerosols as hair, body, room, shoe or car sprays, as well as in paints and in agricultural applications, is superfluous, and they can be replaced by deodorant roll-ons, creams, pumped pressure, nitrogen, or propane–butane mixtures. The United States and Scandinavia have switched to hydrocarbon (HC) propellants for 90 percent of these aerosols.

[1]Based on information provided by the Chemical Manufacturers' Association (CMA 1987a; 1987b).

For insulation, polystyrene foams can be replaced by fiberglass, mineral wool, and cellulose in building applications. In refrigeration equipment, evacuated panel technology can replace blown insulation at great energy efficiency gain (see below). Nonurethane foam can be blown with pentane. High-efficiency air conditioning compressors can use low vapor-pressure hydrocarbons.

The gasoline additive ethylene bromide (CH_2BrCH_2Br; see also Table 1.1) is superfluous if lead-free gasoline is used. The displacement of lead and ethylene bromide should, however, be done in conjunction with the introduction of three-way catalytic converters to avoid emission of benzene and toluene, both of which are carcinogens.

Allegedly, there are no good substitutes available yet for the fire extinguishants halon 1301 $CBrF_3$) and halon 1211 (CF_2BrCl). Methyl bromide (CH_3Br), which is used as a soil fumigant, could be displaced by biological pest control and other changes in agricultural practices.

Table 3.3 shows that CFC-11 and CFC-12 are by far the most important CFCs.[2]

Technical Substitution Potentials. Miller and Mintzer (1986) investigated the short-term reduction potential at a cost of less than $5 per pound in applications of these CFCs. They found that for CFC-11, possible reductions are as follows: foam blowing agent, 50 percent; aerosols, 90 percent; refrigerants, 25 percent; and other uses, 25 percent. For CFC-12, possible reductions are refrigerants, 25 percent; aerosols, 90 percent; foam blowing, 50 percent; and other uses, 25 percent.

It appears that much larger reductions than those identified by Miller and Mintzer under cost considerations are technically feasible. A recent analysis by Makhijani et al. (1988) found that all but two applications of CFCs and halons could be replaced with available technology. Recently announced plans by several governments to achieve a 90 percent or greater phaseout by the mid-1990s (see below) are further indications of these possibilities.

The Montreal Protocol. In September 1987, representatives of 43 United Nations member countries signed a so-called U.N. Protocol in Montreal on the control of chemicals that deplete the stratospheric ozone layer of the earth's atmosphere. This protocol sets out reduction

---

[2]Note, however, that other CFCs that are currently less important already contribute about a quarter of the warming commitment from all CFCs (see Section D in Chapter 2).

targets for CFCs and halons. The chemicals controlled by the protocol fall into two groups (Dechema 1987) and are listed in Table 3.3:

Group I: CFC-11, 12, 113, 114, 115
Group II: halons 1211, 1301, 2402

After ratification of the protocol, the following control measures are to go into effect:

- A production and import freeze on the basis of 1986 figures six months after ratification but no later than 1990.
- A 20 percent production cut by 1992.
- A further 30 percent production cut to be introduced by the end of 1999.
- Review and revision, if needed, every four years starting in 1990.

It took about a decade to negotiate this agreement. In the meantime, further research has revealed that a major speedup of the Montreal reduction targets is urgently needed.

The Need for Accelerated Phaseout of CFCs. A study by the U.S. Environmental Protection Agency (Hoffman and Gibbs 1988) found that stabilizing the ozone layer basically means stabilizing atmospheric chlorine and bromine levels. But under the Montreal Protocol, concentrations of these chemicals would still rise substantially and inexorably, due to the long atmospheric residence times of the chemicals. Even if it is assumed that each and every nation will participate in the Montreal agreement, the ozone layer would continue to be destroyed. The EPA study concludes that halting the depletion of the ozone layer will require a virtually 100 percent elimination by 1998 of the CFCs regulated under the protocol, compared to a 50 percent cut under the current agreement. It will also require a freeze of methyl chloroform production at current levels.

National and Local Responses. National or regional initiatives to accelerate the Montreal schedule are warranted. Fortunately, the new scientific evidence is beginning to create an international political response. During the 120-nation London conference on protection of the ozone layer in March 1989, the 12 nations of the European Economic Community (EEC) announced that they would phase out all protocol substances by the year 2000.

Previously, various countries had already begun to take steps to speed up the Montreal schedule. For example, the FRG government has reported a pledge by its aerosol industry to reduce its annual aerosol production of 26,000 tons in 1986 to about 5000 tons by the end of 1989 (Toepfer 1988). This pledge amounts to a 40 percent reduction of the total 1986 CFC production in the FRG of 53,500 tons (about 5.4 percent of the world total) ten years ahead of the U.N. Protocol schedule.

The most far-reaching national response so far has come from Norway and Sweden (Enquête-Kommission 1988). In Norway, a 50 percent cut of ozone-depleting substances is planned for 1991, followed by a 90 percent cut by 1995. The production and importation of CFC-based foams, dry cleaning agents, and solvents will be banned by that year. In Sweden, the government has set a target of 50 percent reduction by 1990, and a total ban of CFCs outside applications for medicine by 1994.

Local governments also have begun to play an important, if more symbolic role. In the United States, a number of cities have passed ordinances that ban the use of CFC-based styrofoam cups and food containers. In the state of New York and elsewhere, efforts to phase out styrofoam packaging are primarily motivated by the nonbiodegradable nature of such packaging and its impact on landfill requirements. Recycling CFCs from air conditioners and refrigerators has also been investigated by one of these cities, but has proven less amenable to unilateral local-level action. In July of 1989, the city of Riverside, California, passed an ordinance that essentially implements a near-total ban of CFCs within three years.

A number of developing countries also have ratified the Montreal Protocol. Though developing nations account for no more than about 5 percent of world production at this time, major increases can be expected unless a deliberate effort is made to transfer non-CFC technology to their growing economies. Indeed, such international assistance may well be a precondition for achieving the 100 percent participation required to protect the stratospheric ozone layer. China and India have indicated their willingness to participate provided that financial and technical assistance is given for substitution investments.

The need for a fresh approach to technology transfer is illustrated by conventional refrigerator technology. The saturation of refrigerators in the People's Republic of China has been increasing dramatically over the last decade and is projected to reach 60 percent soon. To avoid both the ozone-depleting impact of CFCs used in compressors and foam insulation, and reduce dramatically the greenhouse effect from use of these refrigerators (which is dominated by carbon dioxide released in coal-based electricity generation), it is necessary to increase the energy efficiency of refrigerators drastically and eliminate insulating foam and compressor CFC releases from these units.

An advanced, low-impact refrigerator would look something like this: the use of vacuum panels for insulation could eliminate CFC releases from foam insulation[3] while drastically improving energy efficiency, by about a factor of 4 or more over conventional technology. Greater energy efficiency of the cabinet, in turn, would shrink the needed size of the compressor and the coolant circuit. Recycling would further limit CFC losses from these smaller compressors. However, this advanced refrigerator technology is not likely to be commercialized and made available to developing countries in a timely manner without a concerted effort by the industrialized countries.

Transitional Reduction Targets. In general, the following transitional steps and phaseout policies should be implemented within the next five years:

- Abandon CFC uses in dry cleaning, all foams, throwaway dishes, and sprays.
- As a transitional solution, recycle refrigerant fluids and promote closed-cycle application of industrial solvents where no substitutes are available yet.
- Put a steep tax on CFCs to effect earliest development and commercialization of substitutes.
- Require labeling of all CFC containing or CFC-derived products not yet banned.
- Promulgate CFC-minimizing construction and operating regulations for air conditioning and refrigeration equipment.

Global Reduction Potential. The toleration scenario assumes a 50 percent global reduction by the year 2000, and an about 80 percent reduction in 2100 relative to base-year release rates. In view of the current momentum to revise the Montreal Protocol, the phaseout of CFCs could in reality begin sooner than assumed there, could proceed more steeply, and could even result in close to a 100 percent cut by about 2000. This would eliminate most of the CFC contribution to total 2100 transient warming that remains in the toleration scenario.

[3]Note that the majority of CFC releases from refrigerator use are not from the refrigerant, but from foam insulation.

This potential conservatism could provide a small amount of buffer against less-than-expected successes in achieving the reduction targets for other trace gases.

2. Nitrous Oxide (N_2O)

At about 0.25 percent per year, nitrous oxide exhibits the slowest growth rate in atmospheric concentrations among the five major greenhouse gases. Nevertheless, it is an important driving force in the threat of global warming. Due to its 170-year residence time in the atmosphere (Table 1.1), present release rates will need to be cut by 80–85 percent to stabilize atmospheric concentrations.

As shown in Table 3.4, the nitrous oxide cycle is dominated by natural sources. Nitrous oxide sources are the most uncertain of the five major greenhouse gases. Until recently, the main source from human activity was believed to be fossil fuel combustion, with soil cultivation, fertilizer use, and biomass burning next in importance. Some recent research suggests that fossil fuel combustion contributes

TABLE 3.4 Global Budget for Nitrous Oxide[1-4]

N_2O Sources/Sinks	10^6t N/yr	Range	%
SOURCES			
Natural			
Soils tropical forests	3	2.2–3.7	38
Soils temperate forests	1.1	0.7–1.5	14
Oceans and estuaries	2	1.4–2.6	25
Total natural	6.5	4.3–7.8	81
Man-made			
Fossil fuel combustion	0.2	0.1–0.3	3
Biomass burning	0.2	0.02–0.2	3
Soil fertilization/cultivation	1.1	0.01–2.2	14
Total man-made	1.5	0.3–2.7	19
TOTAL SOURCES	8	4.4–10.5	100
SINKS			
Removal by soils	?	?	
Photolysis in the stratosphere	10	7–13	
Accumulation	3.7	3–4.5	46

[1]Ranges from IPCC (1990). For further estimates, see Crutzen (1983), Logan (1983), WMO (1985), Seiler (1985), Marland (1985), McElroy and Wofsy (1987), Wuebbles and Edmonds (1988), and Lashof and Tirpak (1989).
[2]Estimates of sources and sinks do not necessarily balance.
[3]Estimates with accompanying uncertainty bands apply to ca. 1970–1990.
[4](?) Question mark indicates source or sink could be significant but is not quantified or firmly established.

only about one-third of the previously estimated levels, due to an N_2O-producing reaction occurring in some sample containers (Lashof and Tirpak 1989).

The major control options for nitrous oxide are changes in agricultural practices, reduced consumption of fossil fuels, and reduced biomass burning. The former are discussed in Section C. It is estimated that in the context of a highly efficient global energy system, mainly based on renewables, a reduction of about 80–95 percent could be achieved in 2100.

Overall, it is difficult to reliably identify options for reducing nitrous oxide emissions sufficiently to stabilize atmospheric concentrations. We indicate in Table 3.1 a weighted average range of 40–75 percent for potentially feasible reductions.

3. Nitrogen Oxides

Unlike N_2O, the NO_x cycle is dominated by human activities. NO_x emissions from fossil fuel combustion contribute about half of the worldwide NO_x emissions (Table 3.5). Biomass burning is the principal other source. Again, there are large uncertainties in these proportions, as indicated by the ranges in the table.

Reducing NO_x emissions is desirable for public health reasons, but eliminating NO_x is also needed to reduce greenhouse effects. Lower NO_x emissions make more atmospheric OH radicals available for the chemical reduction of tropospheric O_3. This, in turn, reduces the contribution of tropospheric O_3 to climate warming. Insofar as tropospheric ozone also contributes to forest dieback (see Section D), NO_x control yields further greenhouse prevention benefits.

The principal control options for NO_x are more efficient use of energy and replacement of fossil fuels with energy sources that do not involve combustion with air. In stationary combustion, a variety of modified burner components (DENOX devices) can also be retrofitted at affordable cost. In automobiles, catalytic converters or equivalent technologies should be made the universal standards.

These options illustrate the overlap between acid-rain-oriented policies and climate-stabilization policies. A joint consideration of both policy fields is necessary for other reasons. As pointed out by MacDonald (1985), just controlling one acid rain constituent like sulfur dioxide at the expense of NO_x control could shift atmospheric chemistry and increase the concentrations of radiatively active tropospheric ozone. Comprehensive joint reduction of all radiatively and chemically active trace gases is therefore important.

TABLE 3.5 Global Budget for Nitrogen Oxides[1-4]

$NO_x(NO_2)$ Sources/Sinks	10^6t N/yr	Range	%
SOURCES			
Natural			
Microbial activity in soils	20	20	39
Lightning	5	2–8	10
Input from stratospheric oxidation of N_2O	1	1	2
Total natural	26	23–29	51
Man-made			
Biomass burning	4	2–5	8
Fossil fuel combustion	21	21	41
Soil cultivation/fertilizing	?	?	
Total man-made	25	23–26	49
TOTAL SOURCES	51	46–65	100
SINKS			
Wet deposition	21	10–30	
Dry deposition	17	12–22	
Reaction with OH	13		
TOTAL SINKS	51		

[1] Ranges of sources adapted from IPCC (1990). For other estimates, see Logan (1983), NAS (1984), WMO (1985), Kuhler et al. (1985), Crutzen and Graedel (1986), Singh (1987), and Wuebbles and Edmonds (1987).
[2] Due to large uncertainties, estimates of sources and sinks do not necessarily balance.
[3] Estimates with accompanying error bands apply to ca. 1970–1990.
[4] Question mark indicates source or sink could be significant but is not quantified or firmly established.

In principle, more efficient energy use could bring about most of the desirable emission reductions. However, due to the complexities of combustion chemistry, emissions do not always decrease in full proportion to efficiency improvements. Also, pollution control equipment can often be retrofitted to stationary equipment before it is replaced with more energy-efficient equipment. To achieve the fullest reduction of nitrogen oxides, both efficiency improvements and pollution control technologies should therefore be used. Moreover, by combining both approaches, it is possible to realize a given air pollution reduction target at least economic cost.

Because of the uncertainties surrounding the contribution of tropospheric ozone to climate warming, the toleration scenario does not explicitly consider reductions in ozone and its precursors. However, insofar as nitrogen oxides contribute to tropospheric ozone formation, stabilization of the latter at present concentration levels will require that emissions of nitrogen oxides at least remain constant. In view of growing energy service needs in the developing world, new emissions will have to be offset by tightening emission standards for all stationary and mobile sources and by introducing control technologies in those countries and applications where they have been absent so far.

4. Methane

Among the five major greenhouse gases, methane takes a special position. It is the only GHG dominated by sources from the developing countries; and it is the only gas with a relatively short atmospheric residence time (see Table 1.2a). As a result, required reductions to stabilize atmospheric concentrations are a modest 10-20 percent of current emission levels. Available control options in several areas could significantly exceed this reduction target. This is particularly true for the urban-industrial sources of methane.

The global methane cycle is shown in Table 3.6. As with the nitrogen oxides, our present knowledge is marked by large uncertainty. Again, human activities play a major role, with agricultural sources dominating. Natural gas and mining losses and urban landfills constitute about a third of the total anthropogenic emissions. The major agricultural sources—animal husbandry, biomass burning, and rice cultivation—are discussed in Section C.

Landfills produce methane due to the anaerobic fermentation of paper, food, and garden wastes. Methane from landfills is significant because it is radiatively more active than the carbon dioxide that would be released under aerobic decomposition of organic materials.

Mitigation Options. The principal mitigation option for urban-industrial sources is the reduction of coal and natural gas consumption. Properly design energy policies aimed at phasing out fossil fuel consumption would automatically reduce methane emissions as well (see below for a discussion of interfossil fuel switching). In addition, better control of leaks in the natural gas distribution system and better capture of natural gas in coal seams can be pursued.

Landfill gas can be captured for energy production, thus displacing fossil fuels. A widening range of economically viable packaged cogeneration systems is commercially available to utilize methane from landfills of smaller size. Recycling of paper and the composting of food and garden wastes can reduce methane emissions as well.

TABLE 3.6   Global Budget for Methane[1-3]

CH₄ Sources/Sinks	10⁶t CH₄/yr	Range	%
SOURCES			
Natural			
Wetlands (swamps, bogs, including tundra)	115	100–200	22
Termites (and other insects)	40	10–100	8
Oceans	10	5–20	2
Enteric fermentation (wild animals)	5	2–8	1
Freshwaters	5	1–25	1
CH₄ Hydrate destabilization	5	0–100	1
Total natural	180		34
Man–made			
Enteric Fermentation (cattle, etc.)	75	65–100	14
Biomass burning	40	20–80	8
Rice paddies	110	25–170	21
Landfills	40	20–70	8
Gas drilling, venting, transmission and distribution	45	25–50	9
Coal mining	35	17–50	7
Total man–made	345		66
TOTAL SOURCES	525	400–600	100
SINKS			
Reaction with OH in atmosphere	500	400–600	
Uptake by microorganisms in soils	30	15–45	
Accumulation in atmosphere	44	40–48	8

[1]Data as shown from IPCC (1990). For other estimates, compare Bingermer and Crutzen (1987), Cicerone and Shetter (1981), Khalil and Rasmussen (1983; 1985), Seiler (1984; 1985), Ehhalt (1985), WMO (1985), Kuhler et al. (1985), Edmonds and Marland (1986), Fraser et al. (1986), Crutzen (1986; 1987), Crutzen et al. (1986), McElroy and Wofsy (1987), Mathews and Fung (1987), and Wuebbles and Edmonds (1988).
[2]Due to large uncertainties, estimates of sources and sinks do not necessarily balance.
[3]Estimates with accompanying error bands apply to ca. 1970–1990.

Methane Impacts of Interfossil Fuel Switching. A complication related to methane control arises from the option of fuel switching, that is, switching from carbon-intensive coal and oil to less carbon-intensive gas.[4] As shown in Table 1.2a, methane is about 25–32 times more potent as a greenhouse gas than carbon dioxide. So long as methane losses from natural gas systems are small, fuel switching will nevertheless bring greenhouse prevention benefits.

This conclusion is reinforced by the fact that coal mining and oil production are themselves associated with significant emissions of

[4]See Table 4.1 for the relative carbon intensities of the three fossil fuels.

methane. The offset of carbon benefits due to natural gas losses is thus smaller than indicated by the losses in natural gas. Another factor that works in the same direction is the greater efficiency of gas combustion technology compared to coal combustion technology.

Worldwide natural gas lost due to venting and leakage from gas wells, refineries, and transmission and distribution systems is about 1-2 percent of total production, but systems in some countries are leakier.[5] Given these proportions, it is safe to assume that fuel switching at the point of energy production and conversion (e.g., power plants) will definitely preserve a net warming advantage. So will fuel switching at the point of end use (gas-fired furnaces, appliances/air conditioning, packaged cogeneration, etc.), so long as standard levels of system maintenance are observed.

The situation is somewhat different for vehicles running on compressed natural gas. Here, vehicle methane emissions expressed in equivalent CO_2/km driven almost offset the lower carbon intensity of natural gas compared to gasoline (Unnasch et al. 1989). In general, care must be taken that technology choices are assessed on the basis of all greenhouse gas emissions, not just CO_2, and on the basis of delivered energy services.

Global Reduction Potential. We assume that in the longer term, 50-70 percent of landfill methane emissions can be eliminated. We further assume that in the context of a general phaseout of fossil fuels, natural gas flaring, gas system leaks, and mining losses could be reduced in proportion or eliminated entirely.

The weighted average of all long-term estimated reduction potentials combined is 52-62 percent. Atmospheric methane concentrations in the toleration scenario in 2100 are 10 percent above 1980s levels (see Chapter 2), implying that the reduction potentials identified in Table 3.1 are not fully implemented in the scenario. As in the case of CFCs, this reserve could easily be needed to compensate for surprises elsewhere, such as less-than-targeted reductions in other trace-gas releases or stronger-than-expected biogeochemical feedbacks.

5. Carbon Monoxide

The global cycle for carbon monoxide is shown in Table 3.7. Again, anthropogenic sources constitute probably about half the total, but proportions are uncertain.

---

[5]Reportedly, parts of the gas system of the Soviet Union have experienced significant leakage problems.

TABLE 3.7 Global Budget for Carbon Monoxide[1-4]

CO Sources/Sinks	10^6t CO/yr	Range	%
SOURCES			
Natural			
Oxidation of natural hydrocarbons	500	110–170	20
Methane oxidation	500	175–1050	20
Plant emissions	100	50–210	4
Oceans	60	20–90	2
Forest wildfires	30	10–50	1
Total natural	11190	370–2570	48
Man–made			
Forest clearing	500	190–820	20
Energy use	450	350–580	18
Agriculture	250	90–400	10
Oxidation of anthropogenic hydrocarbons	90	0–190	4
Total man–made	11290	630–1980	52
TOTAL SOURCES	2480	1000–4500	100
SINKS			
Reaction with OH	2100	1200–2600	
Soil uptake	?	?	
Accumulation	23		

[1]Ranges adapted from IPCC (1990), Logan et al. (1981), WMO (1985), Kavanaugh (1987), and Wuebbles and Edmonds (1987).
[2]Due to large uncertainties, estimates of sources and sinks do not necessarily balance.
[3]Estimates with accompanying error bands apply to ca. 1970–1990.
[4]Question mark indicates source or sink could be significant but is not quantified or firmly established.

Though carbon monoxide is eventually oxidized to carbon dioxide, it interacts with OH radicals and indirectly influences the concentrations of methane and tropospheric ozone.

As with nitrogen oxides, a reduction in carbon monoxide emissions is desirable for health reasons alone. Again, energy efficiency and fossil fuel substitution could mitigate emissions, but much more stringent source control technologies are also available and will need to be more broadly applied. Automobile and truck emissions are particularly in need of greater control efforts.

6. Carbon Dioxide from Limestone Processing

During cement manufacture, limestone is converted from its mineral form, which is calcium carbonate, to calcium oxide. In the process of this chemical conversion, formerly mineral-bound carbon dioxide is set free. In 1986, this source contributed 136 million tons of fossil carbon dioxide releases to the atmosphere, or 2.45 percent of total fossil emissions (Marland et al. 1988).

Building materials other than cement are widely used, notably in the developing world. They include mud bricks, adobe, poles, and other local materials. However, some bricks, notably those made from carbonaceous clays, cause carbon dioxide releases themselves. Substituting timber for cement and concrete structures could be an option in some regions and countries. Timber construction is widely practiced in the United States and Canada, but timber is much less abundant in many other countries.

Where feasible, this substitution would have three climate benefits: it would not only avoid limestone emissions, but also reduce the fossil fuel requirements of construction materials. In construction, the energy embedded in materials dominates the energy requirement of building activities. Timber processing is far less energy-intensive than cement production. Finally, if harvested on a sustainable basis, timber not only avoids limestone carbon emissions, but also stores atmospheric carbon that was sequestered during growth for a period of several decades, thus providing a temporary sink of carbon (see also Section E).

Besides substituting building materials, greater efficiency of materials use could be important, including improved design practice.

While cement production in the most advanced industrialized countries has leveled off, the potential for major increases in consumption exists in the developing world.[6] However, the magnitudes of carbon dioxide involved are relatively small. Unlike fossil fuel releases, they could be offset by relatively modest reforestation programs.

C. GREENHOUSE GASES AND AGRICULTURE

Human use and alteration of the terrestrial ecosystems contribute to the total release of greenhouse gases with emissions of radiatively active methane (CH_4), nitrous oxide (N_2O), and carbon dioxide, as well as other nitrogen oxides (NO_x) and carbon monoxide, which are chemically active. In combination, the three radiatively active biospheric sources may contribute as much as 20–40 percent of the total warming in the toleration scenario (see Table 3.1).

[6]Under plausible assumptions, cement releases could double.

Methane is produced in cattle breeding, rice plantations, landfill disposal of organic wastes, and biomass burning. Nitrous oxide and nitrogen oxides are produced through the use of artificial fertilizers and other soil cultivation methods, and in the combustion of biomass. Carbon monoxide is also produced in biomass burning. Carbon dioxide is released through land-use changes and agricultural activities including, but not limited to, the direct burning of biomass.

Mitigating these releases requires a major overhaul of current agricultural production methods. These reforms must be seen in a broad context. In this section, we look at such reforms mainly from within the agricultural sector itself. From this perspective, the following overall criteria apply. The world community must

- Greatly increase agricultural productivity in developing countries to satisfy the basic nutritional needs of a growing world population.
- Limit agricultural activity to land areas that can be sustainably farmed.
- Limit agricultural productivity increases to levels that can be achieved and maintained over the long term with sustainable farming systems.

Of course, a stable climate is a key ingredient in any farming that is sustainable. Traditionally, climate stability has been taken for granted. If we now explicitly add climatic criteria, agricultural practices must

- Minimize the impingement of agriculture on carbon pools in soils and forests.
- Minimize the release of greenhouse gases from agricultural activities themselves.

In this section, we investigate the extent to which measures needed to abate GHG emissions can also put agriculture on a sustainable footing in general.

In Sections D and F, we consider agriculture from a broader perspective that integrates all land uses (cropland, pastures, managed forests, etc.) and their impact on carbon storage in forests and soils.

1. Biomass Burning

Biomass burning in connection with agricultural production comprises the combustion of organic waste matter in the field, slash-and-burn shifting cultivation, fuelwood use, and land clearing through forest burning in the course of extending settled areas. Biomass burning can contribute significantly to the global budget of several major trace gases in the atmosphere, notably N_2O and NO_x, CO, CH_4, and CO_2 (see Tables 3.1 and 3.4–3.7).

Most of the biomass burning in agriculture occurs in the tropics and the savanna areas during the dry season. An estimated 600 million hectares of savannah and bushlands are affected each year (Crutzen et al. 1979). Recent NASA satellite photographs suggest that land clearing for development projects and settlements in the Amazon basin has accelerated enormously. A 1 million km^2 cloud of white smoke was found hanging over the area in 1988. Slash-and-burn subsistence agriculture affects a comparatively small 22–30 million ha of tropical forests each year (Houghton et al. 1987). Also, large amounts of agricultural wastes are burned on the field all over the world.

A further source of greenhouse gases is the burning of wood, dung, and other materials for fuel use, much of which occurs in simple stoves and open fires in developing countries. However, the amount of biomass burning in the field, that is, without providing energy services, is probably at least as much and possibly twice as much.[7]

In biomass burning, one has to distinguish between the production of carbon dioxide and the production of other trace gases. To the extent that biomass burning is fed by deforestation, this contributes a net flux of carbon to the atmosphere. If deforestation were compensated for by reforestation or afforestation,[8] this carbon flux would be suppressed. But the releases of the other combustion product trace gases would still remain. It is these other emissions we are concerned with in this section.

An important aspect of biomass burning is that these tropical emissions take place in a photochemically very active and atmospherically dynamic region, in which considerable transfer of tropospheric air to the stratosphere occurs. This makes the large-scale production of chemically active trace gases such as nitrogen oxides and carbon mon-

---

[7]Total biomass fuel use currently is about 2 Terawatt-years per year (see Table 5.1), which is equivalent to about 1.6 billion tons of carbon releases in the form of CO_2. By comparison, Crutzen et al. (1979) estimate that anywhere from 1.4 to 2.9 btC are released annually from biomass burning in savannahs, in shifting cultivation, and on permanent croplands.

[8]Afforestation refers to planting trees on previously nonforestland, whereas reforestation means planting on land previously forested. We also use the term afforestation for planting trees on cropland, though most cropland was forested at one time. The term *net reforestation rate* is used for the difference between reforestation and afforestation rates and deforestation rates.

oxide especially worrisome. Unlike NO_x emissions from fossil fuel use, emissions from biomass burning are not controlled by emission reduction technologies.

A certain amount of biomass burning is part of the natural ecological maintenance cycle. Likewise, traditional shifting cultivation and other indigenous practices do not, per se, represent an ecological threat. If a 10–15-year fallow period is observed, such cultivation is compatible with a sustainable secondary forest cover. It is mainly the shortening of this fallow cycle due to population pressures and social factors such as landholding inequities that is causing ecological havoc. With the wholesale land-clearing for development projects and new settlements, rates of forest clearing and biomass burning have skyrocketed, pushing them far beyond those caused by traditional levels of shifting cultivation (see also Section D).

Mitigation Options. The mitigation of carbon dioxide releases is discussed in detail in Sections D and E. Here, we concentrate on carbon monoxide, methane, and nitrogen oxides. These emissions could be reduced by

- Replacing slash-and-burn shifting cultivation with permaculture techniques (see Section E).
- Using trees felled during land clearing for timber and biomass fuel applications.
- Using agricultural wastes for composting and other forms of soil regeneration, or biogas production, rather than burning them.[9]
- Controlling other forest, savanna, and bushland burning.
- Using improved technologies in the combustion of biomass for energy purposes.

In changing combustion techniques, the trade-offs between CH_4, NO_x, and CO emissions must, of course, be carefully considered. These need to be more comprehensively researched. However, the now dominant uncontrolled combustion of biomass represents the most unfavorable conditions for trace-gas control. A significant trace-gas reduction can be achieved by switching from open fires to controlled burning in more efficient stoves. An even greater reduction could

---

[9]Note, however, that a good deal of in-field crop residue burning is done to suppress crop pests (e.g., maize borers). In these instances, reduced burning will depend on the availability of substitute chemical or biological pest-control methods.

potentially be achieved by first converting biomass into biogas using gasifiers and digesters, and then burning the resulting gaseous fuels with efficient gas cook plates, gas lanterns, and other clean-burning end-use devices.[10] In commercial and industrial applications, wood gasification, producer gas technologies, and integral units with highly efficient gas turbines could find many applications.

The widespread introduction of more efficient fuelwood stoves and biomass gasification technologies is imperative if the basic needs of Third World rural people are to be effectively addressed. This is one of a number of instances where climate-stabilization and development goals are convergent and could be brought into harmony.

Global Reduction Potential. We assume that in the long term, most traditional slash-and-burn cultivation and forest clearing will be phased out, and most agricultural waste burning will also be curtailed. Moreover, with improved infrastructures in Third World rural areas, a significant portion of cleared trees that were previously burned could be made available for other uses, such as biomass fuel or timber applications (Leach and Mearns 1988). We envision that in the context of the highly efficient energy systems that will be needed to reign in fossil carbon dioxide emissions

- The total volume of biomass burning (energy- and nonenergy-related) could be reduced substantially below present levels.
- Most biomass burning would occur under controlled conditions in modern biomass-based energy conversion devices rather than under uncontrolled conditions in the field or in low-efficiency devices.

We estimate, under these conditions, nitrous oxide emissions from biomass burning could be reduced by 50–75 percent, and methane emissions by 20–60 percent (see Table 3.1).[11]

2. Livestock Production

Livestock production contributes about a third of the total methane from human agricultural activities (Tables 3.1 and 3.6). CH_4 in animal

---

[10]As in the case of the expanded use of natural gas, methane leaks must be minimized to assure a net greenhouse benefit (see Section B.4).

[11]A lower reduction potential is assigned to methane in order to take account of possible offsets from biogas technologies.

husbandry is produced by the anaerobic fermentation of organic matter in the rumen and lower gut of animals. The emission rate depends on the type of animal.

A recent review and update of animal methane sources is found in Crutzen et al. (1986). By far the most important animal methane source is the world's cattle, which are estimated to release 65–100 million tons per year (see Table 3.6 and 3.8). According to FAO (Food and Agriculture Organization of the United Nations) production yearbook statistics, the total world cattle population in the mid-1980s was about 1.3 billion, compared to 1.1 billion sheep, 0.8 billion pigs, and 0.5 billion goats. The cattle population in industrialized countries is about 47 percent of the world total, compared to a little more than 20 percent share in world human population.

With a share of 75 percent, cattle dominate methane emissions (Table 3.8). This is mainly because of the large amount of feed and fodder they consume, but also because the fraction of the feed converted to methane (the methane yield) is greater than in other animals.

TABLE 3.8 Methane Releases from Animal Sources by World Region

Animal type and region	Population early 1980's (million)	CH_4 production per individual (kg/year)	Total CH_4 Production (million tons/year)	Fraction
Cattle	1226	45	54.3	0.72
ICs	573	55	31.5	0.42
DCs	653	35	22.8	0.30
Sheep	1138	6	6.9	0.09
ICs	400	8	3.2	0.04
DCs	738	5	3.7	0.05
Pigs	774	1.2	0.9	0.01
Goats	476	5	2.4	0.03
Buffalos	142	50	6.2	0.08
Horses	64	18	1.2	0.01
Mules, Asses	54	10	0.5	0.01
Wild animals	200		2	0.02
Total*			75	1.00

Source: Crutzen et al. (1986).

ICs = industrialized countries; DCs = developing countries.
*Rounding errors.

In cattle, CH_4 yields range from 5 percent (high-quality feed and level) to 9 percent (low-quality feed and quantity) of the gross energy intake. In nonruminant pigs, they range from 0.5 to 2 percent, with an average of about 1.3 percent.

The methane yield also depends on the quality and the amount of feed. The amount of feed is highest for dairy cows, which typically receive three times their maintenance level feed. About 10–15 percent of the world's cattle are dairy cows. The mean CH_4 production by cattle in industrialized countries is about 55 kg/animal/year. Cattle in Third World countries produce only 35 kg/animal/year, partly because of lower-quality feed, partly because of less feed input. Also, many Third World animals are kept for draft-power purposes rather than just meat and dairy production. The world average CH_4 release is 45 kg per head of cattle annually (Table 3.8).

As shown in Table 3.8, total methane production is about equally large from animals in developing and developed regions.[12] Animal herds have grown steadily over the last 100 years, driven by increases in human population. For example, cattle herds were about 310 million in 1890, 640 million by the 1920s, and 740 million around the time of World War II.

Mitigation Options. The following measures could be envisioned:

- Capturing CH_4 released by animals in feedlots or enclosed quarters.
- Converting manure to biogas.
- Breeding for more productive dairy cows, breeding herds, and draft animals, especially in the Third World.
- Shifting meat consumption to less methane-intensive animals.
- Developing feed and fodder conducive to low CH_4 emissions.
- Reducing cattle populations.
- Developing less CH_4-producing rumen bacteria cultures.

The contribution biogas digesting of manure could make to methane reduction is not clear. Some (unknown but probably modest) fraction of cattle manure is now being fermented anerobically under natural conditions without methane capture (e.g., in ponds or in rice-paddy fields). Biogas digesters would increase the fraction of manure

---

[12]The world's herds of goats, camels, buffalos, and mules and asses are almost entirely found in the Third World.

fermented anaerobically, but under controlled conditions. Biogas digesters contribute to methane control and can also provide significant benefits in other areas, such as reduced fuelwood demand, reduced nitrogen oxide and carbon monoxide emissions from open biomass burning, and reduced use of fossil fuels in fertilizer production and rural energy applications.

Producing more meat and milk products per animal is an important mitigation option. In industrialized countries, livestock herds peaked around 1975, and greater productivity allowed for a slight reduction of herds to meet demand. In the Third World, most animals are managed extensively by peasant farmers in family herds, and do not benefit from breeding techniques. Extension services aimed at this deficit could greatly reduce the need for increased herds as populations grow in these regions.

Another important approach to animal methane control is to limit cattle herds and moderate the consumption of beef. Here, one has to clearly distinguish between industrialized countries and developing countries. Per capita meat consumption is currently six times higher in the industrialized countries than in the developing world (78 kg/cap-yr compared to 14 kg/cap-yr). Moreover, while industrial country per capita consumption has risen by another 20 percent in the last 15 years, it has stagnated in the Third World (UNEP 1987).

Reducing Beef Consumption in the Industrialized Countries. Reduced beef consumption in the industrialized countries may be a realistic option because people in these nations consume several times more animal protein per capita than the minimum of about 30 grams per day recommended for a balanced nonvegetarian diet. In fact, people in OECD (Organization for Economic Cooperation and Development) countries consume much more beef and milk fat per capita than medical research on cholesterol and heart disease suggests is advisable. The unavoidable fat intake per unit of protein from meat is 10 to 30 times higher than the fat intake per unit of protein from grain, for example, in bread (Bechmann 1987).

To achieve substantial reductions in the methane releases of industrialized areas from domestic animals, it would not be necessary to switch to a vegetarian diet. Consider the following example: in the FRG, people consumed about 90 kg of meat per capita per year in 1984–1985, according to government statistics. Of the FRG total consumption, 10 percent was in the form of chicken, 25 percent in the form of beef, and the other 65 percent in the form of pork. There were about 15.7 million cattle, of which an unusually large fraction (about a third) was kept for milk production. Though cattle supplied only 25 percent of meat consumption, 75 percent of the total feed and fodder was consumed by them (Bechmann 1987). Cattle also accounted for about three-quarters of total methane releases from husbandry animals.

Reducing the per capita consumption of beef by 50 percent would still allow ample supplies of dairy products while reducing methane production by about 40 percent. The cut in per capita meat consumption overall would be no more than 12 percent. If the beef were replaced by pork, the methane reduction would still be lowered by virtually the same amount, due to the low methane production in pigs. Meat consumption would, in this case, not have to drop at all.

Reversing current meat consumption trends could be pursued using a number of policies. One would be to reform the farm subsidies that now promote overproduction and overconsumption of dairy products ("butter mountains"). A more far-reaching measure would be to charge a climate tax on beef consumption and dairy products. The tax could be recycled into incentive payments to farmers for switching to alternative agricultural production patterns and soil-conserving farming methods (see also Chapter 7).

Consumers might be receptive to better education about the health risks of high meat consumption. In the United States, red meat consumption has already slightly declined in favor of poultry consumption. Consumer groups also point out that most beef (and other meat as well) is heavily polluted with antibiotics, sedatives, and a large arsenal of other pharmaceuticals, as well as with hormones for fast growth. After slaughter, beef is commonly treated with preservatives, antioxidants, coloring agents, emulsifiers, tenderizers, and so on. Such meat also contains residues from chemicals, hormones, and pesticides.

From a climatic point of view, the ban of growth-producing hormones or other treatments can somewhat increase methane emissions, by decreasing the amount of product obtained per head of cattle. Consumer demands for protection from chemicals in the food chain are important and should be translated into new regulations, as happened in the European Economic Community. At the same time, it is important that such regulation be accompanied in the future by measures that somewhat reduce the consumption of beef and dairy products.

Finally, reducing beef consumption in the industrialized countries would be an important component of promoting a more equitable international economic order. Because of the large amounts of grain required to produce beef, the geographic location of cattle herds can be misleading. Most industrial countries do not have sufficient agricultural land to support their meat consumption. Beef production is particularly land-intensive, because 1 calorie of meat production requires 3 calories of grain inputs for pork and 10 calories for beef. Land requirements can be up to 50 times higher than for protein production from grain.

As a result, a great deal of the feed consumed in industrialized countries is not produced on the home farm, but purchased from developing countries. For example, Western Europe imports more than 40 percent, or 21 million tons per year, of its feed grains from the Third World (Bechmann 1987). In addition, about two-thirds of the total domestic grain crop goes to feedlots (Mueller-Elze and Bach 1985).

The agribusiness production of grains for foreign-exchange-earning exports to the industrialized regions is one among several factors in the displacement of the rural poor in the Third World onto marginal, ecologically sensitive land. The magnitude of the food value involved in this trade is significant: the 500 million people suffering from starvation in the Third World could find relief from this condition if they had the cash to buy the grains exported to industrial-country feedlots (see also Section E). In that sense, the present level of meat consumption in the wealthy industrialized countries is directly related to starvation in the poor countries of the world.

In Latin America, beef production for export is often based on tropical forest clearing. Fast-food restaurants are an important buyer of beef from these regions' "hamburger farms." The consumption of beef in OECD countries thus also has a direct link with tropical forest loss (see also Section E). Citizen groups have used boycotts of fast-food restaurants to discourage this form of tropical forest destruction.

Cattle Herds in the Developing Countries. In the developing world, cattle and other animals perform a much broader set of functions than just meat or dairy production. One is as a stabilizing agent of "savings account" to buffer against the year-to-year fluctuations of harvests and crop availability. Often, the animals scavenge vegetative matter in forests and crop residues. They provide fertilizer or fuel from dung, as well as hides, bones, hair, meat, and a renewable source of motive power for plowing, threshing, and irrigation. They are also an integral part of the social and cultural system of traditional rural societies. Because of this integral function within Third World agricultural systems, there is less room for cattle population reduction. On the other hand, with growing populations, the need for animals will increase proportionately.

Overall, some 250 million draft animals are kept in the Third World, compared to 650 million cattle (see Table 3.8). These draft animals alone account for about 30 percent of the total Third World animal methane. Until the power-providing functions of these animals can be replaced by mechanized devices running on renewable forms of energy, there is little room for reducing this methane source.

Income growth will also lead to greater animal protein consumption, though much of this demand could be met from less methane-intensive animals. Third World fish resources could also provide a significant portion of future animal protein needs in developing countries. Unfortunately, large amounts of fish caught in Third World waters are now fed to house pets and animals in the industrialized countries.

Global Reduction Potential. A global limit to beef consumption is not only important to reduce methane releases. It would also free up land for reforestation to allow more carbon storage in the terrestrial biosphere, as further discussed in Section E. However, the global methane reductions from cattle as indicated in Table 3.1 would be difficult to achieve by reduced beef consumption alone. A 50 percent reduction in beef consumption in the industrialized countries would just about offset the population-related growth of methane releases in the Third World, assuming unchanging low per capita beef consumption there. Only in the extreme case of limiting cattle herds to the levels required for dairy production could most animal methane be eliminated.[13] In view of these constraints, it is estimated that animal methane releases in 2100 could be reduced by 20 percent relative to present levels (Table 3.1).

3. Soil Cultivation and Fertilizer Use

Soil cultivation (tillage and fertilizing) in conventional agriculture leads to the release of nitrogen oxides, carbon dioxide, and—in the

---

[13]Currently, about 15 percent of the world's cattle are dairy cows. Assuming some amount of nondairy cattle needed for herd regeneration, constant global per capita consumption of dairy products, and a doubling of Third World populations, the reduction limit would be in the neighborhood of 70 percent.

case of rice cultivation—methane. N_2O is emitted due to the cultivation of natural soils and due to the application of fertilizers (Table 3.4). It is produced during nitrification of ammonium and ammonium-producing fertilizers under aerobic soil conditions. It is also released by denitrification of nitrate under anaerobic conditions, including rice cultivation (Bremner and Blackmer 1978). The contamination of ground and surface waters with fertilizer runoff is a further, less understood source of nitrous oxide releases. Even soil cultivation without the use of artificial fertilizers appears to increase emissions of N_2O and NO_x from microbial activity in the soils.

There are significant (at least one order of magnitude) differences among nitrogen fertilizers (urea, ammonium nitrate, ammonium sulfate, ammonium phosphate, and nitrogen solutions) in terms of the amount of applied nitrogen that is released as N_2O.

Mitigation Options. The following measures can be envisaged:

- Using the lowest-emitting nitrogen fertilizers whenever suitable.
- Applying fertilizer efficiently and in forms that enhance utilization (deep placement of fertilizer, timing of application for maximum effectiveness, water management, timed release, etc.).
- Adding nitrification inhibitors to fertilizers.
- Limiting fertilizer use on the basis of water-quality standards (monitoring groundwater for salination).
- Reducing urea from oversupply of liquid manure.
- Combining application of organic and chemical fertilizers.
- Extending no-tillage or low-tillage farming.
- Covering fields with green plants or mulch.
- Using N-fixing plants such as alfalfa.

Impacts of Organic Farming Practices. The latter four options listed above broadly represent the displacement of "modern" chemical or green revolution agriculture by innovative methods of regenerative or ecofarming and its many variants. While such techniques would allow large reductions in nitrogen fertilizer use, it is important to consider their impacts on yields and cropland requirements. These are discussed in Section C.6.

Implications of some practices for nitrous oxide emissions are also unclear at this time. For example, nitrous oxide emissions from fields under nitrogen-fixing legumes may be the same as those under fertilized crops. The same may be true for untilled soils, though experimental evidence in either case is not conclusive at this point (Groffman and Hendrix 1987).

While the nitrous oxides benefits or disbenefits of no-tillage or low-tillage farming and legume fertilizing techniques are unclear, these techniques offer significant greenhouse gas benefits in other areas. One is the reduced consumption of fossil energy in the production of chemical fertilizers. Another is the reduction of fossil fuels for tractors in low-tillage agriculture. A third benefit is the conservation of soil carbon in humus.

Soil humus is a reservoir for organic carbon. The carbon storage in the soil depends on the soil moisture budget and the type of base rock, the supply of detritus from litter and roots, and the type of soil cultivation (e.g., no tillage versus tilling, and frequency and depth of tilling). Intensive chemical-based cultivation accelerates the loss of soil humus and its carbon content, thus contributing to climate warming (Hampicke and Bach 1980).

Under conventional chemical-dependent agriculture, less humus is actually formed than is destroyed. For instance, in 1985, soil loss in the United States amounted to a staggering 22 tons of soil matter per hectare (Global 2000 1981). This loss rate is large compared to soil formation rates of only 12.5 t/ha in deep soils and 2.5 t/ha in shallow soils. Similar soil losses are induced by agriculture in the Third World, including some forms of subsistence farming and pastoralism (see Section E).

Global Reduction Potential. The processes driving nitrous oxide releases are so little understood that the reduction potentials shown in Table 3.1 are purely speculative at this point, and may be too optimistic. We estimate that in the longer term, releases of nitrous oxide emissions from soil cultivation could be reduced by a modest margin. If organic fertilization techniques should emit significantly less N_2O than chemical fertilizers, large reductions could be possible. The issue of increased fertilizer requirements for food production is discussed in Section C.6.

4. Rice-Paddy Fields

The world's rice-growing fields are a major source of methane releases to the atmosphere. From measurements taken in Andalusia, Spain, Seiler et al. (1984) deduced methane release rates ranging between 2 and 14 mg/m²/h. There were strong seasonal variations with maximum values at the end of the flowering stage, and minimum values during the tilling stage and shortly before harvest.

Methane production is limited to wetland (paddy) production. In dryland production, no methane is produced. At this time, most rice production is in wetlands. Anaerobic fermentation of organic crop residues is an important factor in total methane releases. Short-stemmed high-yield varieties produce less residue per unit of rice product, but require increased fertilizer inputs. Long-stem medium-yield varieties that can adapt to a great variety of conditions may, however, offer the greatest overall benefits.

The global emissions of 35–59 Mt/yr deduced on the basis of the Spanish findings are somewhat on the low side as compared with the estimates of 25–170 Mt/yr given in Table 3.6.

Mitigation Options. Rice is the staple food especially in the developing countries of Asia, but also in Africa and Latin America. In view of the high population growth rates and the already precarious food situation in these regions a reduction in rice production can hardly be visualized—let alone be justified.

The following possible measures are briefly considered:

- Avoiding uncontrolled anaerobic fermentation of crop residues.
- Adopting organic fertilizing techniques using biogas digester residues.
- Breeding resilient and efficient strands suited for dryland cultivation.
- Increasing rice cultivation under rainfed conditions on dry uplands.

Improved techniques in the use of rice straw for building materials (roofing) could reduce the anaerobic fermentation of long-stem residues. Using straw as feedstock for biogas digesters instead of plowing it into the paddy could help reduce uncontrolled methane emissions.

Plant breeding and selection could have a useful role, if those strands were developed that make optimum use of natural nitrogen fixation while at the same time reducing methane emission. Strands that produce good yields on drylands should also be emphasized. It is estimated that a 20 percent reduction in methane emissions could possibly be obtained from such developments, as shown in Table 3.1.

5. A Broader Perspective: Technological and Social Reforms in Agriculture

The need to reform current agricultural systems is being increasingly recognized, partly because these systems produce huge surpluses and starvation at the same time, and partly because chemical food contamination and ecological damages from groundwater pollution, soil erosion, loss of soil fertility, salination, loss of species diversity, and failing pest resistance have reached major proportions.

The impoverishment of rural social structures, both in the industrialized countries and in the developing world, has also caused growing concern. Its manifestations are the loss of rural jobs, the decline of rural communities, the bankruptcy of many small farmers, and the concentration of farming into ever-larger units. These trends reinforce the negative impacts and potential risks of chemical agriculture, because careful maintenance of the land becomes less practiced or feasible, and large pools of people with experience and direct knowledge from working the land vanish irretrievably into urban life (Jackson 1980).

Fortunately, the need for major reforms of agricultural production methods point in a direction that would reduce climatic impacts as well. The key techniques of regenerative farming or ecofarming are mulching, conservation tillage, planting legumes, use of organic fertilizer, locally self-sufficient production of animal fodder, biological pest control where possible, and maximum utilization of local conditions and natural biotope relationships. All these methods work to reduce losses of carbon stored in soil humus and releases of carbon dioxide from fossil fuel use in agriculture and fertilizer production.

The same principles of ecofarming are being applied in the developing countries, where they are often part of broader experiments in ecodevelopment. Many methods are being used, including the farming systems approach (Redclift 1984; Bull 1982), which concentrates on small holders and seeks to develop increased and more reliable yields of a range of crops and livestock, including tree fruits.

Agroforestry schemes are one variant of this approach (see also Section E). Like its industrial-country equivalents, this practice seeks to make best use of on-farm inputs, such as organic fertilizer, while minimizing the need for expensive off-farm inputs such as irrigation, nonlocal seeds, and chemical fertilizer. At the same time, many different garden, tree, and field crops are cultivated, and synergisms with animal husbandry carefully optimized. Because ecofarming systems can produce food and fuel for basic human needs with less capital requirements and inputs of imported goods, they can offer broad economic and social benefits for developing countries.

Impacts on Yields. The traditional objection to regenerative farming is that it would lower yields and agricultural incomes. Market acceptance problems with organic produce also have been cited. If true, the broad introduction of these practices could make the maintenance of proper nutrition levels more difficult and costly to maintain.

More importantly, there could also be a conflict with climate stabilization itself: if the goal of growing sufficient food for rapidly expanding Third World populations is to be achieved without significant additional trace-gas releases from conversion of forest land into cropland (see Section E), agricultural productivity in most Third World countries has to be greatly improved. Conversely, it can be argued that higher short-term yields in tons per hectare from "modern" chemical agriculture are ultimately an illusion, since they tend to destroy the ecological basis for this fertility irreversibly.

In considering these issues, it is useful to take a look at recent empirical evidence regarding the cost and yields of alternative agricultural methods.

Ecofarming Experience in Industrial Countries. Preliminary results from on-going long-term field test comparisons conducted by the U.S. Department of Agriculture with the Rodale Research Center in Pennsylvania are encouraging. Of a sample of 800 farmers who converted to regenerative farming, 83 percent had equal or better earnings. Yields did drop for a number of crops, but not for a number of others. Moreover, drops in the initial years after conversion were partially or totally compensated for by better results in later years (Brody 1985).

Similar results were obtained from detailed analyses of ecological farming results in the FRG, including a government survey of the country's 1600 biologically oriented farms. Yields were 10–20 percent lower for wheat and potatoes, but just as high as in conventional farms for beets, various fodder crops, and vegetables. Labor inputs were about 10–20 percent higher, but the savings in pesticides, fertilizer, and other inputs turned out to pay for this extra cost and the somewhat lower yields as well. Overall, the West German farms achieved slightly better earnings than conventional farms (Bechmann 1987). It is estimated that due to a variety of factors, yields from regenerative farming would improve at least 10 percent over a 20–30-year period (Bossell et al. 1986).

Bechmann (1987) constructs a scenario in which regenerative farming techniques become the norm for West German agriculture, and a portion of agricultural land is converted into interspersed biotopes that restore both biotic diversity and aesthetically attractive landscapes. He finds that this system, combined with a 10 percent productivity gain, would be as adequate to supply the country's nutritional needs and habits as conventional agriculture is now.

The growing market acceptance of, and active demand for, organically grown food was recently documented by the U.S. *Wall Street Journal* (*WSJ* 1989). Pesticide-tainted fruit has increased public awareness. According to a 1988 Louis Harris poll, 84 percent of U.S. consumers said they would buy organically grown food if it were available, and 49 percent would pay more for it. Sales of organically grown produce and products in the United States doubled between 1983 and 1988 and have reached $1 billion. In 1988, some 20 big U.S. supermarket chains began stocking organic produce. Some restaurant owners have switched to organic food because of what they say is its superior taste.

Both in the FRG and in the United States, the percentage of total agricultural land farmed regeneratively is still very small — less than 1 percent. Nevertheless, large farmers in California and elsewhere, including big suppliers like Sunkist, are beginning to convert some of their production to organic methods. These and many other examples indicate that organic farming is economically viable, but has been hampered by strong disincentives to its practice in existing government subsidy programs. Still, most U.S. organic farmers expect the percentage of organically grown produce to increase to 8–9 percent over the next few years.

Ecofarming in the Third World. Large numbers of peasant farmers in the developing world still rely on traditional methods that are in many ways similar to modern ecofarming. Unfortunately, due to population pressures and the displacement of poor populations onto more fragile lands, methods that formerly were sustainable are being intensified and applied to ecosystems for which they are not adapted. Also, crop yields in traditional forms of subsistence farming are very low, and raising these yields is critical.

One of the most versatile and promising approaches to ecofarming in developing countries is agroforestry. Agroforestry is a whole-system approach that addresses the many needs and capabilities of small and subsistence farmers. In such a system, which can involve several dozens of crops instead of one or two crops, up to a quarter of the land may be kept under trees. Here, a diversity of fruit-bearing and other useful trees are planted in various densities and clusters moving from the rural settlements outward. Trees are also located along field perimeters, pathways, roads, and around houses, serving several purposes at once, including soil protection from wind and water erosion, watershed stabilization, fuelwood, structural timber needs, shading and air conditioning of dwellings, cash income from tree fruits, and a "savings account" for emergencies.

One of the most authoritative reviews of agroforestry experience is found in Leach and Mearns (1988). They emphasize that agroforestry is not a panacea, and is not practical under all circumstances. Nevertheless, the findings on yields from monitored projects hold great hopes for improving agricultural yields and reducing pressures on forests. Crop yields in a large variety of African settings increased from 40 to 90 percent or more, while peasants also were able to harvest wood fuel in excess of their family requirements.

One important component of these improvements in yields is the fertilizing effect of trees when integrated into the farm. In Senegal, the leaves from 50 acacia trees per hectare proved to be equivalent to 50 t/ha of manure. As a result, annual millet production doubled. In addition, the average tree produced enough pods to increase cattle stocking. In other projects, tree leaves helped boost livestock weight gain because they contain four to five times as much protein as grass. This allowed farmers to experiment with "zero-grazing" practices (Leach and Mearns 1988).

As Leach and Mearns point out, the intensification of crop and livestock production by means of agroforestry methods can at the same time reduce the need to clear woodlands and other tree resources. This, in turn, can reduce biomass burning (see also Section E).

There also has been good experience with other modern ecofarming in Africa regarding the reduction in artificial fertilizer use (Egger 1978). Results from two comparative studies on maize yields in Kenya and sorghum yields in West Africa are shown in Figure 3.1. A comparison of maize production on the basis of green revolution, inefficient traditional, and improved traditional practices shows that good husbandry and use of local seeds can substantially exceed the results from hybrid seeds and chemical fertilizers applied with bad husbandry, and can reach at least 60 percent of the yield obtained with hybrid seeds, fertilizer, and good husbandry (Leach 1982). In view of the accelerated degradation of soils that usually accompanies chemical agriculture, the long-term difference in yields may be even smaller.

Similar results are reported by Wolf (1986), who reviewed results

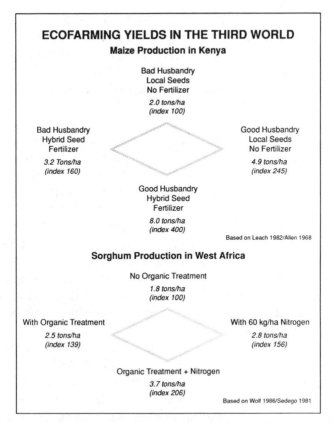

Figure 3.1 *Grain production yields from ecofarming practices in Africa compared to green-revolution agriculture yields.*

from semiarid West Africa. In field studies there, organic fertilizer techniques increased yields of sorghum from traditional methods by almost 50 percent, reaching 90 percent of artificial fertilizer yields. When artificial fertilizers were applied in combination with straw, composting, and manure, yield grew by another 20–30 percent above the one reached with chemical fertilizer alone.

The partial substitution of artificial fertilizers and other capital-intensive green revolution inputs with organic sources and ecofarming practices would also help reduce the foreign-exchange requirements of many developing countries, and the upfront capital needed by peasant farmers to increase their productivity.

These results suggest that regenerative farming could be greatly expanded both in industrialized and developing countries without negative consequences for the goal of increasing Third World agricultural yields. On the contrary, without a conversion to such ecologically sound approaches, the loss of arable land, notably in the tropics, threatens to accelerate beyond control (see also Sections D and E). Nor does one technique have to be used at the rigid exclusion of the other. Substantial field experimentation is needed to see how both approaches can be combined in an optimal way under varying local conditions.

What about Bioengineering? Given the crisis of growing starvation and limited food supplies, the advent of genetic engineering has raised hopes that the world might find a "high-tech" solution to the food problem. If so, the same technology could also offer solutions in the search for greenhouse gas abatement. After the breeding of the green-revolution hybrid varieties, a second generation of bioengineered plants might emerge that combines the high yields and good fertilizer response of the green revolution varieties with the resilience of traditional varieties and with new features such as low fertilizing needs, low methane emissions in rice, and better pest and drought resilience. Similarly, methane production in cattle rumen might be controllable with bioengineered bacteria.

At the same time, there is justified concern that bioengineering intervention in the natural gene pools might backfire. Given past experience with human "improvements" of nature and with controlling the side effects of technologies, it is difficult to see why bioengineering will be free from similar negative consequences. In view of the unprecedented scale of human intervention represented by this new technology, and the scale of runaway biological pollution that could ensure, the risks of negative ecological impact are magnified.

But these dangers aside, genetic engineering may prove to be handicapped by the same social power relations that made its predecessor, the green revolution a mixed success, at best (Wolf 1986). This could be especially true in the Third World, for which it would ostensibly be most beneficial.

Because green-revolution agriculture was for the most part introduced without also implementing land reforms and other social policies, it further concentrated landholdings in developing countries. But worse, it reconstituted the entire natural environment of the Third World's rural poor (Agarwal 1985). In effect, the poor's traditional skills and intuitive knowledge of their environment were rendered useless on a massive scale, with no replacement of an equivalent means of subsistence. In many cases, the new agriculture robbed them of scarce natural resources like fodder, agricultural waste, and so forth, that they traditionally had access to (Banerjee and Kothari 1985). It worsened the split between a small prosperous class of modernized farmers and a swollen underclass of subsistence farmers and landless laborers.

While the former now participate in a system that salinates soils with irrigation practices made possible through large-scale development schemes, pollutes groundwater with nitrates and pesticides, and impoverishes species diversity and soil humus, the latter are pushed into the environmental frontier, where they destroy fragile soils and forests by virtue of their sheer numbers, and with traditional subsistence practices that are no longer adapted to their locales.

Though great progress in good production was made through green-revolution methods, these successes arguably rest on unsustainable foundations. Furthermore, much of the fruits of this progress are exported to industrialized countries. What remains within the Third World is so inequitably distributed that mass starvation has actually increased. Meanwhile, modern agriculture also has added to the foreign-exchange and debt burden of the developing world.

Bioengineering could probably provide some valuable contributions to food production, if carefully controlled and applied in limited measure. The difficulty in realizing this potential lies in the need to not only meld appropriate technology with high technology, but also to join the two technology cultures in a viable development process. To perform this feat requires special policies and equitable social structures to foster and protect the participation of local people.

It is difficult to see how the introduction of high-tech, bioengineered animals and plants, along with a new wave of foreign experts and corporate marketing efforts, will bring agricultural productivity gains within reach of the capital-poor Third World smallholder. Like the green revolution, genetic engineering as a technology brings with it a whole set of institutional and social relationships that is likely to reinforce the Third World's dependence on imports of technology and know-how, and increase their foreign-exchange requirements.

Reversing the social polarization and ecological damage that accompanied the green revolution, while retaining and absorbing its positive elements into a sustainable, regenerative farming practice, will require, above all, a massive social mobilization of the Third World poor. This can only be done with a capital-efficient, self-reliant approach, with land reform and the construction of advanced agricultural knowledge by local communities themselves at its core. Such knowledge will have to build on the farmers' own informal "research system" (Redclift 1984) and the indigenous technical knowledge of their environment (ITK approach). It also requires extension services based on local social institutions, with the ability to reach out to groups unfamiliar with the written word.

There are already precedents where these approaches have proven successful, but the broad-based diffusion of such forms of participatory, sustainable development is being hindered by many constraints (Leach and Mearns 1988), including the grossly inequitable patterns of land tenure typical of most developing countries.

D. CARBON RELEASES FROM DEFORESTATION AND SOIL DESTRUCTION: HOW SIGNIFICANT ARE THEY IN CLIMATE WARMING?

1. Overview of the Global Carbon Cycle

An important area of uncertainty in the greenhouse problem is the net flux of carbon dioxide into the atmosphere due to changes or disturbances of natural life systems. To appreciate the relative roles of fossil fuel consumption and carbon releases from the biosphere, it is helpful to take a closer look at the natural carbon cycle of the world. These main proportions of the natural carbon cycle are shown graphically in Figure 3.2. Table 3.9 summarizes the key sources and sinks. As indicated by the range of values given in the graph and in the text below, there are still major gaps in our knowledge of the carbon cycle. A detailed discussion can be found in Trabalka and Reichle (1986).

Though carbon makes up about 4 percent of the earth's mass, most of it is contained in inorganic rock material or enclosed organic material. Only a tiny fraction (0.04 percent) participates in the atmospheric–biological–oceanic carbon cycle that influences the world's climate. This cycle occurs between five major reservoirs: the atmosphere itself, the terrestrial biosphere consisting of land biota and soils, the upper (mixed) layer of the ocean, and the deep ocean (Fig. 3.2).

Most (about 95 percent) of the carbon participating in this cycle is contained in dissolved form in the deep ocean. In principle, the deep ocean could buffer against disturbances in the other carbon-cycle reservoirs, and restore the atmosphere close to its original condition. However, the deep ocean is separated from the mixed layers of the upper ocean (0–75 m) by a thermocline. The rate of exchange across this boundary is very slow and, like so many aspects of the carbon cycle, not precisely known, but probably more important than the net

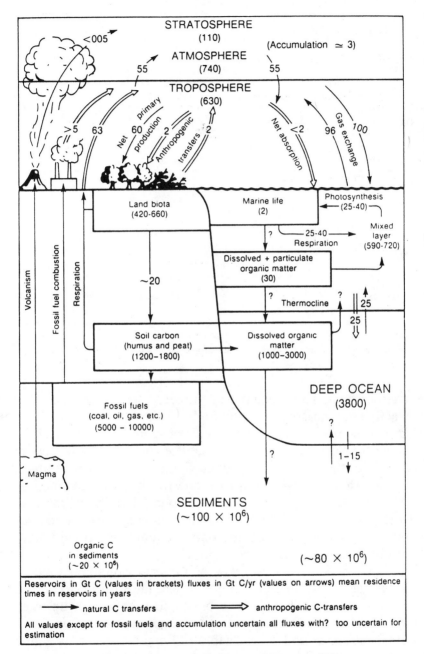

Figure 3.2 *The global carbon cycle, mid to late 1980s.*

TABLE 3.9   Global Budget for Carbon Dioxide[1]

CO$_2$ Sources/Sinks	10^{9}t CO$_2$/yr	10^{9}t C/yr	%
SOURCES			
Natural			
Gross annual release from ocean	381(376–390)	104	60
Gross annual release from land	216(32–440)	59	36
Total natural	597(408–830)	163	96
Man-made			
Fossil fuel use	18(16– 20)	5	3
Land use conversion	7(0– 10)	2	1
Total man-made	25(16– 30)	7	4
TOTAL SOURCES	622(424–860)	170	100
SINKS			
Ocean			
Gross annual uptake	392(388–396)	107	63
Land			
Net primary production	220(183–257)	60	35
Accumulation	10(10–15)	3	1
TOTAL SINKS	622(581–668)	170	100

Source: Trabalka (1985) and Wuebbles and Edmonds (1988).
[1]The estimates with the accompanying error bands apply to any year between 1970 and 1987.

terrestrial fluxes. To the extent that the deep ocean does act as a buffer, this activity is measured in thousands of years.

On a time scale less than thousands of years, the world's climate is thus controlled by the exchange processes in the carbon-cycle reservoirs. Here, the *terrestrial systems* of vegetation and soils are more important than is suggested by their relative size. The *land biota* alone account for 420–660 billion tons of carbon (Houghton 1986). About 90 percent of the carbon in vegetation is contained in the woody biomass of forests and trees, and only the remaining small fraction is found in crops and grasslands.

But an even larger terrestrial carbon reservoir of about 1500 btC is found in the humus materials of the surface layer (top 1 m) of soils. In combination, the carbon in land biota and humus adds up to about 2000–2500 btC. Soils also contain inorganic carbonates on the order of 1000 btC. Peat and other "subfossil" organic materials contain a further 1000–3000 btC. Because they turn over much more slowly, they are not considered further here.

The atmosphere contains presently about 740 btC, which is only a third as much as that stored in biomass. The carbon stored in the upper (0–75-m) layer of the ocean is about the same, or 600–700 btC (Olson et al. 1985). The sum of terrestrial, atmospheric, and upper ocean reservoirs represent about 3300–4000 btC. The terrestrial systems alone contain about 75 percent of the fast-cycle biospheric carbon that is not in the atmosphere already (Fig. 3.2). Note that these biospheric totals are of the same order of magnitude as that stored in the world's conventional fossil fuel resources, estimated at 3800–4200 btC (Rotty and Masters 1985; see also Chapter 4).

With the exception of some of the soil components, carbon stored in these reservoirs turns over rapidly (fast-cycle reservoirs). The typical residence times range from 1 to 30 years. As a result, the natural fluxes of carbon to and from the atmosphere are large, on the order of 60–130 btC/yr. Naturally occurring oscillations in the fluxes of terrestrial systems due to the seasonal change from photosynthesis to respiration in forests outside the tropics are large enough to show up clearly in corresponding oscillations of atmospheric carbon dioxide concentrations in the Mauna Loa monitoring records.

Annual Releases of Fossil and Biospheric Carbon. Currently, net annual terrestrial fluxes of biospheric carbon into the atmosphere are estimated as 1.8 btC/yr for the years around 1980, with an uncertainty ranging from 1.0 to 2.6 btC (Houghton et al. 1987). These fluxes are only on the order of 1 to 2 percent of the naturally occurring fluxes caused by photosynthesis, respiration, and other processes, which explains some of the difficulty in accurately measuring them.

The current annual release of carbon from fossil fuels (about 5.3 btC in 1985–1986; see also Chapter 4) are about three times as great as biospheric ones. But this dominance of fossil fuel releases over biospheric releases probably began only after World War II. Fossil carbon releases did not reach 1 btC/yr until the 1920s. The point value of the current biospheric range, 1.8 btC/yr, was not reached by fossil fuel consumption until 1952, and the high value, 2.6 btC, not until 1960 (Rotty and Masters 1985).

Cumulative Releases of Fossil and Biospheric Carbon. Cumulative net releases of carbon into the atmosphere due to land-use changes and fossil fuel consumption are of significant magnitude, both compared to natural fluxes and compared to the total carbon stored in the fast-cycle reservoirs. Between 1860 and 1985, an estimated 90–180 btC have been released into the atmosphere due to human land-use activities (Houghton et al. 1987). This corresponds to about 5–10 percent of the terrestrial biospheric pool.

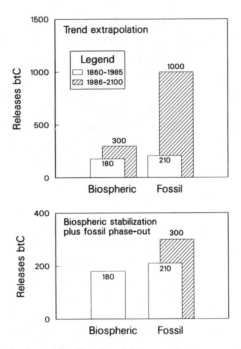

Figure 3.3 Biospheric and fossil carbon releases under two scenarios (1860–2100).

Another 140–180 btC were released from fossil fuels (Rotty and Masters 1985).[14] This is about 5 percent of the total fast-cycle carbon in the upper ocean, the atmosphere, and on land. Figure 3.3 shows the relative proportions based on the upper bound of estimated releases.

Looking into the future to the year 2100, Figure 3.3 shows that cumulative carbon releases from the biosphere could grow by 170 percent above the approximate current value of 180 btC, assuming only constant release rates at the high end (2.6 btC/yr) of current estimates. But this increase in biospheric releases is overshadowed by the trend-based cumulative fossil fuel releases, which would reach, at a 2 percent annually compounded growth rate, about five times their current level. This would shift the balance between cumulative biospheric and fossil releases from about 50:50 between 1860 and 1980 to 30:70 between 1860 and 2100.

Figure 3.3 also shows the relative proportions of carbon releases in the toleration scenario of Chapter 2, in which the biosphere is stabilized at about 1980 carbon storage levels, and fossil fuels are phased

---

[14]These figures refer to 1980. By 1985, the upper bound of the range was 210 btC in round numbers, as shown in Figure 3.3.

out. Here again, cumulative fossil fuel releases would become clearly dominant over time, gaining a share of 70–75 percent of the total 1860–2100 figure.

However, if the destruction of biospheric pools were to further accelerate, this could set free as much carbon as trend-based fossil fuel consumption. The carbon pool in terrestrial ecosystems alone is about three to four times the size of cumulative fossil releases to date. It is precisely these important terrestrial life systems that are being extensively and dramatically altered by human civilization.

In summary, this overview leads to an important conclusion:

• Changes in the terrestrial life systems and soils caused by human activity could pose a large additional climate threat, compounding the impacts of fossil fuels by a significant margin.

2. Biospheric Carbon Releases and Land-Use Changes

Below, the various components of human activity that are changing terrestrial biospheric carbon storage are examined. Here, it is important to consider both direct human impacts due to changes in land use, and indirect environmental impacts on the biological productivity of undisturbed and altered ecosystems. The impacts of human activity on the carbon content of soils and standing biomass are also somewhat different in the tropics and in the temperate or high latitude (boreal) zones.

The greatest change in the terrestrial biospheric carbon reservoir occurs through human agricultural activity. This activity can be broadly defined as the transformation of forests into nonforestland for pasture or crop production purposes. The harvesting of forests for timber in managed forests or industrial plantations also has an impact on stored carbon.

Of course, the reverse process of forest recovery is also at work, since some agricultural land is continually being taken out of production. The reasons may be agricultural overproduction, soil conservation or reforestation programs, the fallow cycle in tropical shifting cultivation, or simply the abandonment of land that has been degraded to the point of nonproductivity.

However, the natural recovery of such abandoned agricultural land may be severely hampered or entirely blocked if overly intensive use has set into motion the process of desertification. This is increasingly the case in many semiarid regions. Even in the humid regions of the tropics, desertification can occur where laterite soils are involved. The loss of arable land then leads to more clearing of forests to maintain agricultural production.

Quantitative Relationships. A quantitative understanding of these impacts on terrestrial carbon storage and releases can be gleaned from the data presented in Table 3.10.

When forests are converted to nonforest areas, there is a drastic drop in the carbon content stored in standing biomass, and some drop in soil carbon. As shown in the table, converting forests into cultivated land leads to a loss of anywhere from half to three-quarters of total stored carbon per hectare. The impact of vegetation clearing alone is most severe in tropical rain forests, where about two-thirds of the total

TABLE 3.10 Carbon in Vegetation and Soils for Different Land Uses and Regions[1]

Land-use type	C in vegetation tons/ha	C in soil tons/ha	Total C tons/ha	Total C Index 1	Total C Index 2
Northern (boreal) forest					
natural	90	206	296	1.00	1.00
managed	72	185	257	0.87	0.87
"recovered"	68	185	253	0.85	0.85
Temperate evergreen forest					
natural	160	134	294	1.00	0.99
managed	120	120	240	0.82	0.81
"recovered"	120	120	240	0.82	0.81
Temperate deciduous forest					
natural	135	134	269	1.00	0.91
managed	100	120	220	0.82	0.74
"recovered"	100	120	220	0.82	0.74
Temperate grassland	7	189	196	1.00	0.66
Temperate woodland	27	69	96	0.49	0.32
Temperate cultivated land					
formerly boreal	5	155	160	1.00	0.54
form. evergr. or decid.	5	101	106	0.66	0.36
Tropical moist forest					
natural (a)	176-250	100-120	276-370	1.00	0.93-1.25
(b)	73-140	100-120	173-260	0.63-0.70	0.58-0.88
mature fallow	70-90	90-108	160-198	0.54-0.58	0.54-0.67
Tropical dry woodland	27-90	50-69	77-159	0.21-0.58	0.26-0.54
(b)	15-40		84-96	0.23-0.35	0.28-0.35
Tropical pasture	5	75-90	80-95	0.26-0.29	0.27-0.32
Tropical cropland	5	70-84	75-89	0.24-0.27	0.25-0.30

Source: Houghton et al. (1987).
[1]Ranges for tropical forests reflect differences between Latin America, Africa, and Asia.
(a)Values from Brown and Lugo (1982).
(b)Values from Brown and Lugo (1984).

carbon are found in the vegetation,[15] and least severe in the boreal regions, where proportions are about the inverse.

Note the croplands and pastures store only very little carbon in their vegetation, on the order of 3 – 7 percent of the soil-plus-vegetation total (Table 3.10). In this respect, they are not much different from natural grasslands and woodlands. Note further that if croplands and pastures are degraded through mismanagement, they can release once again as much carbon from the soil as the original forest clearing that created them released from vegetation. Even before this complete desertification is reached, overly intensive use can reduce the humus content of soils enough to cut soil storage by as much as 40 – 50 percent. Soil loss due to wind erosion also accelerates the oxidation of soil carbon. Similarly, the table shows that the destruction of savannas and dry woodlands due to overgrazing and excessive wood-fuel harvesting can release as much carbon per hectare as the clearing of forests in forested regions.

Even where forests are harvested on a renewable basis, there is a carbon loss of anywhere from 10 to 25 percent in temperate and boreal forests. The same reduced carbon storage is found in "recovered" forests that regrew on abandoned agricultural land.

The difference is even larger in tropical rain forests. In tropical shifting cultivation, a mature fallow area may contain less than 60 percent of the undisturbed carbon storage.

Global History of Land-Use Changes. Before the advent of agricultural civilizations, about 10,000 years before present, forests and open woodlands covered an estimated 6.2 billion ha (Mathews 1983), or about 47 percent of the world's land area. Human activity began to have an impact on forest cover early on, as witnessed by the desertification of the Middle East, but it did not begin to change dramatically on a global scale until about 1800, when world population began its steep increase.

Today, only about two-thirds, or 4.1 billion ha, of the original forest cover is left. If degraded tropical areas supporting shrubs and forest fallows are included, the total comes to 5.2 billion ha. Relative to the original state of affairs, the reduction in forest cover is thus at least

---

[15]We show in Table 3.10 the high estimates for the carbon content of tropical forest vegetation, which are derived from direct measurement. Significantly lower numbers (by 30 – 60 percent) are obtained from FAO/UNEP statistics on growing stocks (Brown and Lugo 1982; 1984).

TABLE 3.11   World Land Area by Category and Use, 1982–1984

	LDCs billion ha	ICs billion ha	World billion ha	World Index 1	World Index 2
Total forests	**3.17**	**2.00**	**5.16**	**0.39**	**1.00**
before human disturbance			*6.2*	*0.47*	
Closed forests	1.34	1.49	2.83	0.22	0.55
Open forests	0.80	0.52	1.30	0.10	0.25
Shrubland and forest fallow	1.03	0	1.03	0.08	0.31
Cropland	**0.81**	**0.66**	**1.47**	**0.11**	0.32
Cropland in dryland areas	0.46	0.24	0.70	0.05	0.15
of which irrigated	0.10	0.03	0.13	0.01	0.03
of which desertified	*0.03*	*0.009*	*0.04*	*0.00*	*0.01*
of which rainfed	0.37	0.2	0.57	0.04	0.12
of which desertified	*0.28*	*0.07*	*0.35*	*0.03*	*0.08*
Pasture	**1.89**	**1.26**	**3.15**	**0.24**	0.68
Pasture in dryland areas	1.53	1.03	2.56	0.20	0.55
of which desertified	*1.23*	*0.38*	*1.61*	*0.12*	*0.35*
Total agricultural	**2.70**	**1.92**	**4.62**	**0.35**	**1.00**
productive drylands	1.99	1.27	3.26	0.25	0.71
of which desertified	*1.54*	*0.46*	*2.00*	*0.15*	*0.43*
Total other			**3.32**	**0.25**	**1.00**
Total land area			**13.08**	**1.00**	

Source: Desertification data from UNEP (1987), historic forestation from Mathews (1983), forest-cover estimates from Postel and Heise (1988), and other from WRI/IIED (1987).

LDCs = less developed countries

16 percent, and could be twice as much on a forest-quality basis. These and other statistics on present-day land use are summarized in Table 3.11.

In the course of this conversion, a large but difficult-to-estimate amount of biospheric carbon was released into the atmosphere. In the Holocene period before agricultural disturbance, long-lived plant material contained an estimated 340 btC more in vegetation, and possibly several hundred more btC in soil humus (Olson et al. 1985). But the speed of change until about 1800 was apparently not large enough to significantly alter the atmospheric equilibrium.

In the period from 1860 to 1980, on the other hand, the rate of biospheric carbon releases due to land-use changes accelerated by one to two orders of magnitude.[16] In that period, the cultivated area world-

---

[16]Assuming that most releases before 1800 occurred within the preceding 2000 years, the average release rate was perhaps (100–500)/2000 = 0.05–0.25 btC/yr, compared to 1.3–1.5 btC/yr for 1860–1980 (see below).

wide increased dramatically, from less than 600 million ha to about 1500 million ha (Houghton 1986). Houghton estimates that this resulted in a release of about 160–180 btC in the same period.

This accelerating growth put enormous pressure on forest resources. Of the total carbon released, between 60 and 90 percent was due to the clearing of forests for permanent agriculture. Reduced fallow periods, increasing land areas under shifting cultivation, and land degradation due to intensified cropland cultivation account for most of the remainder (Houghton 1986).

Biospheric Releases from Temperate and Boreal Ecosystems. The percentage of the biospheric carbon stored in terrestrial ecosystems outside tropical forests is about 320–360 btC, or about 60–75 percent of the global stock (Houghton et al. 1985). These figures illustrate the critical influence of forest maintenance in the industrialized regions on global climate stabilization.

The temperate and boreal ecosystems are mainly located in China, the Soviet Union, North America, Europe, and most of Oceania. These regions contain about 40 percent of the forested land of the world today. The largest share of forestland in these regions is found in the Soviet Union (about 20 percent), followed by North America (about 13 percent). Europe and China contain about 3 percent each.

On the surface, the picture of biospheric carbon dioxide releases for these nontropical regions is assuring. The expansion of agricultural areas had largely ceased in the industrialized regions by the middle of the twentieth century, and forested areas have remained fairly stable. In Europe, they even increased due to the abandonment of significant areas of the less productive agricultural land. Reflecting these dynamics, Houghton et al. (1987) estimate a net biospheric release of only 0.1 btC/yr from boreal and temperate ecosystems in 1980, or no more than 4–10 percent of the estimated world total.

There are, however, geographic areas that do less well. Both the eastern Soviet Union and China are setting free more carbon through cropland increases than they sequester through afforestation and cropland abandonment. In view of the enormous population concentration in China, it is a major achievement that deforestation has been greatly slowed, but the additional population pressures of the next few decades will pose a challenge to the gains made so far, and make further improvements more difficult to achieve.

In the United States, growth in grain exports and urban sprawl have led to a reduction of forested areas by about 10 percent since 1960

(WRI/IIED 1987). Continuing large soil losses in U.S. agriculture also are a factor, accelerating the oxidation of soil carbon (see Section C).

In this respect, the U.S. soil conservation program begun in 1985, which sets aside 16 million ha of highly erodible cropland for planting in trees or grass, is an important step. It could render North American terrestrial ecosystems a net sink for carbon (Postel and Heise 1988). However, nonsustainable logging practices, notably clear-cutting in erosion-prone forestlands, continue in the United States and elsewhere at large scale. A significant portion of forestlands in industrialized countries are understocked or degraded.

Acid Rain Impacts. There is other cause for concern that the relatively stable overall picture for the nontropical areas might change. The above carbon release estimate does not take into account the massive and sudden forest damage recently detected in Europe and the eastern United States, caused by acid rain and other pollution stress. In 1986, some 52 and 36 percent of the total forest area in the FRG and Switzerland, respectively, had been damaged (Giesen 1986). In central and northern Europe as a whole, more than a fifth of forested area has already been damaged. This area alone corresponds to more than 7 percent of the nontropical forest area. Together with affected areas in the United States, some 10 percent of the nontropical forests are likely involved.

The potential magnitude of climatic impacts from this environmental problem are, as Woodwell (1987) and Bach (1985) point out, large and widely underestimated. Suppose the 10 percent of forests already damaged die back. From the figures on total standing biomass quoted above, it follows that this could release as much as 35 btC to the atmosphere. This corresponds to seven years of world fossil fuel consumption at the current level. Depending on the speed of such a process, this dieback alone could greatly increase global biospheric releases. Natural recovery would likely take many decades.

Biospheric Releases from Tropical Ecosystems. Houghton et al. (1987) find that the global net biospheric carbon flux of $1.8 + 0.8$ btC/yr in 1980—about as large as the annual fossil carbon release from total world oil consumption (Chapter 4)—was almost entirely from the tropics. Some 40 percent of the total comes from tropical Latin America, 37 percent from tropical Asia, and 23 percent from tropical Africa.

By far the largest national contributor to Brazil, which alone accounts for 20 percent of the total. According to the same data source, just one of the nine Amazonian provinces of Brazil releases more biospheric carbon than all temperate and boreal regions combined. Eighty percent of the total is accounted for by just 15 countries, with the biggest players, other than Brazil, including Indonesia, Colombia, Thailand, Laos, Nigeria, and Cote d'Ivoire.

The land-use changes and other human activities driving these releases are complex and regionally diverse.[17] *Forest clearing for cropland* is by far the major driving force. The intensification of slash-and-burn subsistence shifting cultivation also plays a major role. *Fuelwood collection*, including selective cutting of trees for charcoal production, can have devastating consequences, particularly around cities. Notably in Latin America, *cattle ranching* for beef exports drives the clearing of large tracts for pastures. *Export of tropical hardwoods* to cash-rich Northern markets also is a major factor, partly because of its indirect effects: it leads to the construction of logging roads, which brings in massive numbers of colonists in its wake. In some countries, like Brazil and Indonesia, *colonization programs* are additionally promoted by governments and even international development agency grants. Similarly, the World Bank and bilateral aid agencies support large *hydro dam projects* that often destroy huge virgin forests. *Acid rain damage* is also being detected in several Third World areas.

Tropical Deforestation. Around the turn of the century, about 2 billion hectares of tropical forests existed. By 1980, only about half of this total was left. Current net deforestation rates in the tropics are large. The most widely accepted current estimates are from a joint study by the Food and Agriculture Organization (FAO) and the United Nations Environment Program (UNEP), from FAO Production Yearbooks, and from work by Norman Myers. Recently, these estimates were brought into comparable form and synthesized by Houghton et al. (1987). The range of the FAO/UNEP and Myers numbers is shown in Table 3.12.

The deforestation rates shown in the table include closed tropical rain forests in humid regions where the dense canopy prevents grass from growing on the forest floor; open forests including woodlands in semiarid regions; and forest fallow and dry shrub lands. The net loss in total forest area is about 11–14 million ha/yr, equivalent to the size of Great Britain. The greatest area losses occur in the open forests and shrub lands of the seasonal and dryland stocks. Clear-cutting in tropical rain forests is 4–7 million ha, but about two-thirds is on mature

---

[17]See Postel and Heise (1988) for further discussion.

TABLE 3.12 Annual Rates of Tropical Deforestation and Desertification

	Area by climate region			
	Moist million ha/yr	Seasonal million ha/yr	Dry million ha/yr	Total million ha/yr
Forest clear-cutting				
(1) f. perm. pasture & cropland	0.2-1.0	0.7-2.8	2	4.2-5.8
(2) f. shifting cultivation				
(a) gross	4.1-6.7	6.4-11.9	5.9-6.4	16.5-25.0
(b) mature fallow	2.8-4.4	4.4-7.5	4.1-4.4	11.3-16.7
(c) primary (net)	1.3-2.3	2.0-4.0	1.8-2.0	5.2-8.3
Total gross (1+2a)	5.1-6.9	9.2-12.6	7.9-8.4	22.3-29.2
Total net (1+2c)	**2.3-2.5**	**2.1-4.2**	**3.8-4.0**	**11.0-14.1**
percent of stock	*0.2*	*0.25-0.5*	*0.4*	*0.36-0.46*
Forest degradation				
primary forests				4.3
Grand Total				**15.3-18.5**
percent of stock				*0.5-0.6*
Global loss of arable land				**5-7**
percent of stock				*0.33-0.47*

Source: FAO (1980), FAO/UNEP (1981), and Houghton et al. (1987).

Notes: (a) Total gross is total clearing for all purposes.
      (b) This is clearing for shifting cultivation on land previously used for that purpose.
      (c) Total net is total gross minus land clearing under (b).

fallow. Clearing of undisturbed tropical rain forests is about 2.3–2.5 million ha/yr (Houghton et al. 1987).

Expressed in percentages, the net deforestation rates range from 0.2 to 0.5 percent of the world's forestland per year. If we combine deforestation and degradation, about 0.5–0.6 percent are affected per year. These figures may seem modest, but they would still mean a loss of 10 percent of forests in less than 20 years.

Moreover, in individual countries, deforestation rates are as much as ten times as high. Since 1966, the Ivory Coast has lost 56 percent of its forest cover, and a score of other nations have lost from 20–40 percent (WRI/IIED 1986; 1987). Recent satellite data indicate that Brazil's deforestation rate has increased more than twofold since 1980. In the Amazon basin, rates of up to 10 percent per year have been reported (Salati 1987). At that rate, about 90 percent of the present 400 million ha of Amazon rain forest will have been destroyed in about 20 years.

By comparison, afforestation in the tropics is lagging far behind. In Africa, on average, only one in about 30 ha cleared is replaced by reforestation or by afforestation elsewhere. About half the African countries have no significant reforestation efforts at all. In tropical America, the ratio is about one in ten, and in Asia about one in five (WRI/IIED 1987).

Forest Degradation. The deforestation rates alone do not accurately measure the extent of the damage to the world's forest stock. They do not include the vast areas that were simply degraded, but not fully cleared of trees. From an ecological point of view, impoverishment is also caused in areas where selective logging is practiced, but surrounding vegetation is damaged, and in areas where plantations diminish species diversity and forest stability. This partial degradation has itself a significant impact on carbon releases. It is estimated that each year 4 million ha of forestland are degraded by these and other human activities (Table 3.12).

Desertification. Desertification is also spreading, leading to the release not only of vegetation carbon, but also of soil carbon. According to estimates by the U.N. Food and Agricultural Organization (FAO 1980), about 5–7 million ha of the world's arable land under cultivation are completely lost for agricultural production every year. For comparison, the FAO projects that about 200 million ha of new arable land will have been added between 1980 and 2000. At current rates of loss of arable land, this addition would little more than compensate for losses due to soil degradation.

The most affected regions are those with semiarid or arid climates (drylands). Of the world's 4.5 billion ha of drylands, 3.3 billion are classified as biologically productive, and are used as pastures, rainfed cropland, and irrigated cropland. Almost half the world's cropland is in dryland areas, and about 80 percent of the world's pastures are found there. But 61 percent of this area is already desertified, that is, production of biomass and soil reproduction are at least 25 percent diminished (WRI/IIED 1986, 1987; see Table 3.11). UNEP estimates that an even larger fraction of this land is at risk of becoming desertified.

As can be seen from Table 3.11 (land categories), most of the desertification occurs in the developing world. Fifty-seven percent of the Third World cropland is in dryland areas, and here, fully two-thirds are already desertified. Of the 20 percent or so under irrigation, a third is affected.

3. Feedbacks of the Greenhouse Effect on Biospheric Carbon Releases

Increased carbon dioxide concentrations and warmer climates could themselves lead to feedback effects on biospheric carbon fluxes to the atmosphere. The present understanding of such possible mechanisms is highly inadequate, especially in view of their enormous implications for greenhouse risks (Lashof 1989).

The Up Side: Increased Photosynthesis? Increased carbon fixation in plants in response to higher carbon dioxide concentrations would be a negative feedback mechanism. Such a compensating effect has been proposed by some, partly as one explanation of the "missing sink" in the current accounting of the global carbon cycle (see Section D.4).

The carbon dioxide levels used to promote plant growth in commercial greenhouses are several times those occurring naturally, not just 25–30 percent higher. On the other hand, chamber experiments show that some plants respond in a linear fashion over a wide range of CO_2 concentrations. There is some uncertainty whether crop plants will respond to increased CO_2 concentrations even more or less so when their access to energy, water, and mineral nutrients is limited, as is the case in the natural environment. A further complicating factor is that weeds and pests are likely to be even more enhanced than crops. The transfer of chamber or hothouse experience to the natural environment could therefore be misleading.

But for CO_2 enhancement of plant growth to explain the "missing sink" in the carbon cycle (see below) it is above all the *forest biomass* that would have to be increasing. Almost all the data on CO_2 enhancement are on crops. A recent analysis of feedback mechanisms in the climate system suggests that an enhancement effect could be real but appears too small to account for the "missing sink" (Lashof 1989).

The Down Side: Increased Respiration? Warmer temperatures brought about by the greenhouse effect could lead to a positive feedback by accelerating the respiration of plants while producing little impact on photosynthesis (Woodwell 1983). When respiration outpaces photosynthesis, trees release more carbon than they fixate, as evidenced by the seasonal fluctuations in the Mauna Loa records. A 10°C increase in temperature can increase plant respiration by anywhere from 30 percent to more than four times. Several degrees of warming are expected in the northern region for one degree of warming in the global average surface temperature (see Chapter I D.1).

Woodwell (1986) warns that if respiration exceeded photosynthesis for an extended period of time, a large-scale dieback in the Northern Hemisphere could occur. If the recent experience with forest damage in Europe is any indication, such forest dieback could occur within a decade once the process is set in motion. By contrast, the migration of species more adapted to the new temperature regime into the destroyed regions would take a century or more. The rate of increase in global warming would be decisive for the extent of the devastation that might occur.

The threat of a positive temperature feedback on biospheric carbon releases is not limited to forests. It would affect all vegetation and soils, especially those of the middle and higher latitudes including the tundra (Billings et al. 1982; 1983; 1984). The forests of western and northern Europe would be among the areas affected in a major way.

Thus, depending on the rate of warming, up to several hundred billion tons of carbon could be released from the biospheric pool in a matter of a few decades. The annual rate of such releases could outstrip releases from fossil fuels for an extended period. The potential consequences of massive forest dieback and further warming illustrate the danger of seeing greenhouse warming as mainly beneficial for Northern Hemisphere regions.

4. The Missing Sink: Is the Biosphere Active or Neutral?

There remains at this time a significant gap in climatologists' and geophysicists' ability to balance the global carbon-cycle budget. Carbon-cycle models built on geochemical knowledge can account for the fossil carbon releases and the atmospheric CO_2 concentrations monitored at Mauna Loa since 1959, but not for an additional biospheric carbon release of 1.8 btC/yr or more, as suggested by Houghton et al. (1987). To bring the models into balance requires a neutral biosphere or a slight net sink in the biosphere, pointing to a $1-2$ btC discrepancy between models and total (biospheric plus fossil) carbon release estimates (Trabalka et al. 1985; Houghton et al. 1987). This balancing requirement of current carbon-cycle models is reflected in the climate-warming calculations of Chapter 2. Consistent with other climate-warming modeling exercises, they treated the biosphere as neutral.

The discrepancy between model requirements and land-use-based estimates has led to speculation about additional carbon sinks not presently accounted for. Various proposals have been advanced, including the possibility that the $25-30$ percent increase in atmospheric CO_2 concentrations since 1860 might somehow have stimulated carbon uptake in plants. As already discussed above, there is no experimental evidence available to support this suggestion. It is urgent to improve the data base on this question.

One difficulty in resolving this issue is that the magnitude of the discrepancy is small compared to the large natural fluxes occurring in the terrestrial–ocean–atmospheric system, and thus difficult to detect. Modeling exercises will have to contend with these carbon-cycle uncertainties for some time. Resolving the question of whether the biosphere will react with positive feedback to climate warming or with negative feedback to increased carbon dioxide concentration will require significant additional research. Ideally, it would include extensive monitoring of actual ecosystems over at least a decade (Trabalka 1985).

In the meantime, this uncertainty means an exposure to significant climatic risk. The entire discrepancy could be an accidental outcome of temporary opposing effects. Once these overlaps disappear, we might find that there is no hidden carbon sink after all, and that the global greenhouse effect is driven by the full sum of net biospheric carbon releases and fossil releases. One should, therefore, distinguish between current modeling conventions and actual risks posed by the known man-made changes in terrestrial ecosystems.

Risk-Minimizing Policy Responses to Carbon-Cycle Uncertainties. Whether it is assumed that net biospheric releases are negligible or not has significant policy implications for climate stabilization. If the current biosphere is treated as a neutral "black box," any additional tree planting would offset fossil releases. If, on the other hand, the biosphere is seen as a significant source of carbon, tree planting will be needed first and foremost to offset biospheric releases from the destruction of forests and soils.

Recall that in our modeling calculations of Chapter 1, climate stabilization in the toleration scenario was achieved under the assumption that the biosphere was neutral. If the biosphere is in fact a carbon source, and the toleration scenario is still to be realized, fossil–biospheric trade-offs would be feasible only after biospheric carbon pools have been stabilized. The stabilization of these pools is itself a major task, as further discussed below.

Treating the biosphere as neutral has been a common assumption in climate-modeling exercises not only because of model calibration difficulties, but also because these exercises focused on scenarios of growing fossil fuel consumption. In that case, biospheric releases do become less important (see Fig. 3.3), though positive feedbacks from forest dieback could still make a large contribution.

But relative proportions of fossil and biospheric releases are different under a scenario of climate stabilization, in which fossil fuel use is largely phased out within the next century. As a result, the "residual" term of biospheric releases, including its uncertainty range, gains in importance (Fig. 3.3). After a few decades, net biospheric releases would even become dominant if they were to continue at present levels.

If climate policies are to minimize risks,

- We must assume that biospheric releases increase the risk of greenhouse warming in direct proportion to the net flux calculated from land-use changes.

This is equivalent to saying that biospheric carbon directly and fully adds to the warming from fossil carbon releases. From the same risk-minimizing perspective,

- Global land-use, land conservation, and afforestation policies should be aimed, at a minimum, to achieve a neutral biosphere close to present levels of terrestrial carbon storage.
- Optimally, these policies should be geared toward increasing terrestrial carbon storage above current levels.

This treatment is warranted in particular because: (1) biospheric carbon releases since about 1860 were as large as those from fossil fuels to date; (2) according to our best knowledge, the current rate of biospheric releases is in fact not zero but substantial; and (3) this biospheric release rate could grow through massive forest dieback due to air pollution and warming stresses.

To translate such a risk-minimizing perspective into policy action, it is necessary to explore in some detail the global opportunities for reforestation and afforestation.

E. RESTORING BIOSPHERIC CARBON POOLS IN FORESTS AND SOILS

1. How Many Trees Should Be Planted? A Climate Perspective

From the perspective of climate stabilization, we need to know how much tree planting would be needed to achieve the following goals:

- Eliminating current net annual *biospheric* carbon releases of about 1–2 btC/yr.
- Compensating for annual *fossil* carbon releases of 5.5 btC/yr.
- Returning terrestrial ecosystem carbon storage to the level that existed before industrialization (i.e., fixing an estimated 90–180 btC of biotic releases).
- Sequestering all (fossil and biospheric) carbon releases since industrialization (an estimated 250–400 btC), plus all future fossil carbon releases that would unavoidably still occur under a rapid fossil fuel phaseout (another 300 btC; see below).

Figure 3.3 shows the relationships of some of the quantities involved.

The first target should be seen as the minimum action required to mitigate against biospheric carbon risks. Essentially, it is geared toward stabilizing the terrestrial carbon pool as closely as possible to current levels. Further destruction of existing carbon pools, now accelerating at an alarming rate, would first be steadied and then reduced in the near to medium term, while tree planting elsewhere would increasingly offset remaining losses. Eventually, a steady-state situation would be achieved. In the course of the transition to this steady state, a significant amount of current terrestrial biospheric carbon would still be shifted to the atmosphere. To bring carbon storage back to current levels, a significant further afforestation effort would have to be undertaken.

Under the second goal, tree planting would be implemented at a sufficiently fast rate and broad scale to maintain current storage and to begin to offset fossil releases as well. This fossil fuel offset, would, however, be only temporary. If one assumes that fossil fuel consumption will continue after the newly planted forests have matured, tree planting becomes mainly a strategy for gaining time; once the added forest areas have matured, carbon sequestering will cease, and fossil releases will be unabated. Naturally, this time element could be very important, both in terms of the climate-stabilization goal of slowing the *rate* of warming, and in terms of the time needed for putting into place highly efficient energy technologies and renewable energy sources.

The third goal overlaps with the second, but is oriented toward minimizing biospheric climate impacts by sequestering historic carbon releases from the atmosphere. This "rollback" approach reflects the climatic risks associated with our lack of certainty about the eventual impacts of these releases. The magnitudes involved are shown in Figure 3.3. As discussed in Section D, net biospheric carbon releases since industrialization were almost as large on a cumulative basis as those from fossil fuel consumption.

Of course, this rollback of past biospheric emissions would still be counteracted by fossil releases. To pursue the maximum reduction of anthropogenic climate risks, one would want to phase out fossil fuel consumption over the next few decades, as assumed in the toleration scenario; sequester all historic carbon releases since the onset of industrialization; and also offset all future fossil and biospheric releases that are inevitable during the completion of the fossil phaseout. This would lead to a return to preindustrial atmospheric conditions and a steady-state situation in which no net anthropogenic carbon releases, either biospheric or fossil, would occur. The magnitudes involved are again shown in Figure 3.3.

An Upper Limit Analysis for Global Tree Planting. Could any of these carbon sequestering targets be realistically approached? In principle, the percentage of the earth's land surface under forests could probably be returned to the 6 billion ha level of preindustrial times. If left alone and given sufficient time, nature might do the job on its own. However, the climate problem must be addressed in the next five to ten decades. This presents a logistic constraint.

Also, there is now 1.5 billion ha of agricultural cropland, much of it on areas formerly forested, and the need for food production will increase with growing populations in the developing countries. Further vexing issues in developing a reforestation scenario are the political and social constraints placed upon implementing the necessary land-use changes.[18]

The discussion below presents a rough sketch of a series of reforestation efforts that could conceivably be pursued over the next 50–100 years. The purpose of this sketch is to estimate the order of magnitude of the biotic carbon-fixing potential under optimistic assumptions. It is not to develop a detailed action plan. Nor does it address all the practical implementation problems associated with tree-planting programs, which are many and daunting. The principal purpose of this exercise is to better define what minimum requirements should be applied for the curtailment of fossil fuel consumption under the goal of

---

[18]For a sophisticated treatment of these constraints based on experience in Africa, see Leach and Mearns (1988).

climate stabilization. These minimum requirements are defined by the maximum contribution other policy areas could make toward this goal.

2. More Forests versus More Food: An Inherent Conflict?

In the past, discussion has focused mainly on potential conflicts between food production and energy crop production (e.g., the use of sugar cane or corn for conversion to fuel alcohol). The climate threat now adds a potentially even larger competitor for agricultural land. How can global afforestation be achieved in the face of growing food demands? And does biomass energy production fit in any more at all?

It is estimated that an additional 200 million ha of land will be devoted to agriculture in the Third World by the year 2000. Other investigations of the world's agricultural future have suggested that the world's cropland area might be doubled in the longer term, adding another 1500 million ha (FAO 1978).

But according to FAO estimates, all but about 11 percent of the world's land area—that is, about as much land as is now under cultivation—offers serious limitations to agriculture. The more land taken under production in the more marginal regions, the greater the likelihood for accelerated loss of cropland from soil degradation and desertification. Deforestation can itself accelerate the loss of soils and cropland, because of watershed damage, lowered water tables, and accelerated wind- and flood-related erosion.

In the near term, further forest clearing for agriculture, and the attendant releases of biospheric carbon, seem therefore unavoidable. However, at least over the long term, increasing agricultural demands on the one hand, and maintenance of forests on the other hand, are not necessarily incompatible. Even a limited amount of biomass energy production, albeit from trees rather than from cropland, could be part of a sustainable pattern. Options for counteracting the pressures of increased cropland demand on forests can be grouped into three categories. Measures in the first category are aimed at reducing the demand for agricultural land and thereby reducing demands on forest resources. They include

- Raising the crop growing and livestock raising productivity of poor, small Third World farmers whose appallingly low yields force them to clear land (see also Section C).
- Replacing slash-and-burn shifting cultivation by agroforestry and tree-crop permaculture systems.

- Making biomass energy use more efficient.
- Recycling paper, timber, and other wood products.
- Limiting infringements from hydro dams and other development projects on tropical forests.
- Avoiding "hamburger farms" in tropical forest areas.
- Reducing beef consumption.

The goal here is to leave as much land as possible under forest. The second category of measures aims at producing more forestry products on existing forest land. It includes

- Improving management of existing forests and plantations.
- Switching to better harvesting practices (harvesting more species, minimizing damage to standing trees, better utilizing total biomass).
- Reforesting and fully stocking degraded forests.

The third category consists of tree planting in a wide variety of settings:

- Integrating trees into agricultural systems.
- Reforesting surplus agricultural land in the industrial regions.
- Afforesting degraded semiarid lands.
- Establishing peri-urban fuelwood plantations around Third World cities.
- Planting trees in cities and along roadways.

Above all, the rate of biospheric degradation in those carbon pools that still exist must be slowed soon. Without such action, reforestation in other areas is like opening the tap to draw more water into a leaky bucket.

Tree-Crop Permaculture and Agroforestry

Tree-crop Permaculture. Like organic farming, this is a form of agriculture that builds on traditional practices. It has particular potential in tropical forest areas, and is widely practiced in the developing world. This form of agriculture mimics the natural forest system while building on the indigenous techniques of the peoples living in and around them. Trees producing various oil seeds, nuts, and grain-equivalent fruits or natural products such as gum or selected species of timber are planted in multistory permaculture systems. These forest systems are then managed to yield food and other products while maintaining a diversified forest cover.

For example, large populations in humid West Africa produce coffee, cocoa, and palm oil as well as many other minor food and other products from tree-crop systems that look to the untutored eye exactly like natural forests. The harvesting of rubber from within natural forests without destroying them is practiced by Brazilian subsistence farmers and is another illustration.

Tree-crop agriculture as a modern practice is less far-fetched than it might sound. The tropics, with their large variety of species, offer many more tree products that are both edible and palatable than the temperate zone forests. Oil palms, breadfruit trees, leafy products of lettucelike quality, and a large variety of nut and fruit trees are examples.

The enormous genetic pool contained in tropical forests also represents an untapped resource for the in-forest cultivation of special purpose plants with potentially very high commercial value. This unique potential of tropical moist forests, accumulated in millions of years of natural evolution and now squandered, could give developing nations a competitive advantage and future export opportunities worth far more than the current income derived from cash crops and unsustainable hamburger farms.

Agroforestry. If tree-crop permaculture takes the farm into the trees, agroforestry brings the trees onto the farm. As such, it combines the maintenance and improvement of soil fertility with increases in carbon storage. Anywhere from 10 to 25 percent of agricultural land may be under trees, but a real competition with crop production and livestock raising is avoided. Fallow enrichment by planting leguminous trees, alley cropping, boundary tree planting, and community and private woodlots are some of the specific techniques used in this approach. In many areas, boundary planting of hedgerows alone could provide half the local fuelwood and fodder needs.[19]

Agroforestry can also make a major contribution to meeting rural fuelwood needs, and offers advantages over large fuelwood plantations. While fuelwood yields per hectare are larger in plantations than in agroforestry, yields per tree are as much as an order of magnitude higher in agroforestry schemes (Leach and Mearns 1988).

---

[19]See Leach and Mearns (1988) for detailed case studies.

Liquid Fuels from Oil-Bearing Trees. Many countries could cultivate oil-bearing trees for obtaining liquid fuels and other petroleum-product substitutes. By using oil-bearing trees, a carbon-storing and soil-protecting tree cover could be maintained. Provided that the biomass fuels are burned with good emission controls, the goal of climate stabilization and biomass energy production would complement each other rather than compete.

This scheme is not without historic precedent. At one time, close to 100,000 people were employed in Brazil in cultivating the babassu tree for vegetable oil production. More than half the fruit of these swamp trees consists of oil, and they yield several hundred nuts per year, equivalent to a barrel of oil per tree or more. During World War II, liquid fuels derived from the Amazonian babassu tree were burned in diesel engines, while the solid fruit residues were converted to coke and charcoal.

Other trees with liquid fuel potentials are the oil palm and the Nypa palm. The latter, which is native to the mangrove forests of Southeast Asia, is reported to yield at least twice as much alcohol per hectare as sugarcane.

Replanting Degraded Semiarid Land. Conflicts between food production needs and reforestation requirements can also be minimized by reforesting desertified and degraded *semiarid regions*.[20] These regions comprise about 1.2 billion ha and can be found throughout the world at the fringes between the tropical and temperate zones. They include the subtropical mediterranean climate zones in northern Africa and southern Europe, the tropical savannas, and the Asian steppes.

While agroforestry has so far been most successful in the humid and subhumid regions, it is technically just as suited for application in semiarid and even arid regions. In fact, such applications provide the most dramatic benefits and improvements. Studies of the impact of acacia tree planting on Sahel zone farms indicate that the land-carrying capacity of the area could be increased by a factor of 3–4 by such planting.[21]

---

[20]Semiarid regions are variously defined as regions receiving more than 200 mm/yr but less than 500 mm/yr in rainfall, typically concentrated in one season. Irrigated farming is needed beyond low-productivity cultivation of a few drought-resistant species. Beyond that, the climate characteristics of semiarid regions are highly variable.

[21]See Leach and Mearns (1988).

A principal problem with agroforestry in semiarid regions is the long (up to 15-year) establishment time for trees. This delay in benefits conflicts with the short-term risk perspective forced on subsistence farmers and pastoralists trying to survive in an unpredictable environment. Providing viable forms of risk reduction is a key challenge for successful development assistance projects in these regions.

Despite the lack of precipitation, many areas are potentially recoverable. There are a number of well-known "rain harvesting" or catchment techniques that allow the funneling of precipitation from a larger surface area to the root system of planted trees. These techniques have been used in Israel, in the Sahel zone, in the U.S. Southwest, and elsewhere.[22] Shelterbelts and livestock exclosure can help maintain and improve biomass stands.

At one time, many semiarid regions supported large tracts of subhumid forestland. Reforestation would be difficult at first, but could in time make semiarid climates less dry. Some 85 percent of the precipitation received on land originates from nonoceanic sources, notably from plant transpiration itself (Winstanley 1983). This feature of the water cycle can thus provide a positive feedback effect once trees have taken hold again.

Woodland afforestation and agroforestry in semiarid regions might initially do no more than compensate for ongoing soil losses and desertification processes on agricultural land elsewhere. In the longer term, it could increase terrestrial carbon storage.

The carbon storage and growth rates that could be achieved on these lands are, of course, smaller than those obtainable from afforestation of subhumid and humid areas. Afforestation also would be more difficult and expensive. But the sheer magnitude of available land area still represents a significant potential for carbon sequestering.

Urban Forestry and Agriculture. Currently, some 2 billion people live in cities worldwide. Their environments generally lack greenery, and the concentration of paved areas and energy use creates urban summer heat islands that raise temperatures several degrees above those found in the countryside. In North America and other industrialized countries where air conditioning represents a major portion of total electricity demand, these heat islands significantly increase fossil carbon releases.

Both problems could greatly benefit from urban tree-planting cam-

[22]For a review of these techniques and of pertinent NGO (nongovernment organization) experience, see Pilarski (1988) and also Leach and Mearns (1988).

paigns. The direct carbon-fixing gains from such planting are modest; if one tree were planted for each urban resident, the total would be equivalent to about 2 million ha of forest, less than one-tenth of 1 percent of the world total.

However, trees can reduce the air conditioning loads in buildings by as much as 50 percent. Akbari et al. (1988) calculate that in U.S. cities, a tree planted for shading and evapotranspiration cooling of the microclimates surrounding buildings eliminates 10–14 times as much atmospheric carbon as a tree planted on forestland. The large difference is due to the avoided electricity consumption, most of which is fossil-fuel-based.

The combined effect of direct sequestering and indirect fossil carbon savings can make planting trees in urban areas one of the most cost-effective climate-stabilization measures available (Krause and Koomey 1989). With growing use of air conditioning in southern European and Third World cities, these benefits will be relevant to increasing numbers of people worldwide.

Tree planting in urban areas could also be part of urban food production. In developing countries, large numbers of urban poor have no access to the cash required to buy healthful, fresh produce. Small-scale, highly productive rooftop and multistory garden farming in cities could meet many food demands locally (Wade 1981). Urban agriculture would reduce the demand for rural land and could avoid portions of the expensive and fossil-fuel-intensive chain of transport, storage, refrigeration, packaging, and processing that characterizes industrialized food systems. The combination of fruit-bearing trees, fuelwood trees, and "air conditioning" trees with urban agriculture could be a promising element in addressing the basic needs and capital shortage crisis of Third World urban populations.

Urban agriculture is not without precedent. For example, in China, at least 85 percent of vegetables consumed by urban residents are grown within municipalities. Two large cities, Shanghai and Peking, are completely self-sufficient in terms of vegetable production. Chinese city dwellers also produce fish, chicken, duck, piglets, and tree crops in sizable quantities (Wade 1981).[23]

Surplus Food Production and Excess Consumption in the Industrialized Regions. Because of productivity gains, the Western industrialized nations are faced with chronic agricultural overproduction, re-

---

[23]Even in some industrialized countries, urban gardens have recently seen a revival. The FRG is an example.

sulting in large grain, butter, and milk surpluses. Notably in Europe, but also elsewhere, these surpluses have already led to the retirement and afforestation of some of the less productive agricultural land. But government subsidies, agribusiness lobbies, and the concern of maintaining a viable base of small farmers and rural jobs have prevented a more complete adjustment. There are many opportunities to return significant additional agricultural land to forest.

Such newly planted forests will reduce the need for agricultural subsidies and will yield economic benefits from timber products. These could at least partially offset the loss of agricultural income. Afforestation would also help stabilize soils in regions of high erosion.

As already discussed in Section C, the high and often excessive beef consumption in OECD countries also leads to excessive agricultural land requirements. In the industrialized world, fully 60 percent of all grains are used to produce meat, compared to 30 percent worldwide (Bechmann 1987). A cut in beef consumption would thus free up a major portion of cropland and pastures.

Since 1970, meat consumption in the industrialized world has increased by 20 percent per capita. A gradual return to about 1970 per capita levels overall, combined with a 50 percent reduction in beef consumption as explored in Section C, would present a significant change in nutritional habit but not any real hardships if done gradually over time. As a benefit, such policies would free up large tracts of OECD land.[24]

The potential for cropland afforestation may grow further as agricultural reforms in the socialist countries progress. Productivity gains from such reforms could easily outstrip the moderate growth of their populations. On the other hand, per capita meat consumption in these nations is lower, so that the percentage of land that would become available for afforestation would be smaller.

Recycling of Wood Products. High-yield industrial plantations fix carbon very rapidly but store much less carbon over time than a natural forest (see Section E.3). To minimize the need for managed forests and industrial plantations, it is important to limit the consumption of timber and tropical hardwoods as well as the consumption of paper and pulp. The average OECD person consumes about as

---

[24]See Bechmann (1987) for a quantitative scenario analysis of these impacts, as applied to the FRG. Note that the 50 percent reduction suggested for beef is larger than that for meat as a whole, due to the 3.3-fold difference in grain inputs required to raise beef and pork, respectively.

much wood in the form of paper as the average Third World person consumes in the form of fuelwood. Recycling of paper is not nearly as widely practiced as it could be, given adequate incentives. Such recycling would at the same time save about 40 percent of the (usually fossil-based) commercial energy inputs during pulp and paper production.

Are Northern Croplands Needed to Feed the Developing World? An objection to this long-term picture of cropland and pasture afforestation might be that food exports from surplus-producing developed countries will be needed to support large and growing populations in the developing countries. Several detailed analyses have shown that this is not the case (George 1976; Moore-Lappé and Collins 1979; Moore-Lappé et al. 1981; Drenham and Hines 1984; Goldsmith 1985; Banerjee and Kothari 1985). They conclude that the current international system of OECD-dominated agricultural trade, and the many assistance programs accompanying that trade, are a primary *cause* of the growing starvation in the Third World, not a solution that should be expanded.

The pressure of foreign debt and the growing integration of Third World agriculture into the world commodity market has led to a major shift of agricultural land from domestic food production into large-scale export crop production dominated by international agribusiness corporations.[25] According to FAO statistics, an estimated 14 percent of Third World agricultural cropland is now devoted to export production. At the same time, Third World self-sufficiency in basic grains, which has already dropped to about 92 percent, is projected to drop to 85 percent by the year 2000. What is more, the grains and other feed exported to industrialized countries from the Third World are almost entirely used for producing meat, that is, for high-income luxury consumption rather than basic needs (see Section C).

The grain imports required to compensate for this shift of agricultural land further undermine the self-reliance of developing countries: grain exports from industrialized nations to Third World countries are routinely subsidized. Such low-cost grains undermine small Third World peasant farmers, while shifting nutritional habits and food preferences toward imported grains such as wheat. Free emergency relief supplies can have a similar detrimental effect. The loss of indigenous food production, in turn, makes developing countries more vulnerable

---

[25]For a detailed discussion of this process in Africa, see, for example, Drenham and Hines (1984).

to bad harvests and more dependent on emergency relief from abroad.[26]

Recently, the direct production of beef in developing countries for export to the wealthy nations has also become significant. Notably in Latin America, cattle ranches or "hamburger farms" (Myers 1981) have been established through the large-scale clearing of tropical forests to provide low-cost beef for northern consumers. This mining of a major climatic resource is yet another thread in the tapestry of food and ecological subsidies provided by the developing countries to the industrialized regions.

Partly because of the extreme internal inequities in the distribution of income in the developing nations, this international agricultural orders leads to the growing use of the most fertile Third World regions for raising those crops that cash-rich OECD consumers buy (coca, coffee, feedgrains, fruit, and meat), while developing countries are becoming more and more dependent on imports of wheat and other grains for feeding their poor populations.

A solution to Third World food shortages thus requires approaches that lessen poverty and income disparities. Some of the most important policy options are debt relief for developing nations by industrialized countries, stimulating food production for the domestic market, and improving agricultural productivity with ecofarming techniques that require less machinery, fossil fuels, and other foreign-exchange and capital intensive inputs, as discussed in Section C.

3. Potential Impacts of a Global Afforestation Campaign: A Scenario Exploration

There is unfortunately no inventory that would classify worldwide land areas by their suitability for afforestation. Short of such a detailed study, a simple rough sketch is developed here.[27]

Table 3.11 shows the world's forest and agricultural areas by current use and by industrialized versus developing regions. For each major land-use category, Table 3.13 indicates the type of afforestation or other land-use conversion envisioned. It also shows the total areas planted, the fraction of land involved in each category, and the average

---

[26]A detailed case study of this process is provided by Andrae and Beckman (1985) for Nigeria.

[27]Among other things, we incorporate proposals by nongovernment organizations (NGOs) involved in tree planting (Pilarski 1988) and perspectives summarized in Leach and Mearns (1988) and Postel and Heise (1988).

TABLE 3.13 Scenario of Land-Use Changes for Sequestering Atmospheric Carbon

Current land use	Area reforested or converted			Conversion to (land-use)	Carbon fixing factors	
	Cumul. million ha	Fraction of total	Rate mill. ha/yr		rate t C/ha-yr [1]	cumulative t C/ha [2]
Industrial countries						
Cropland	100	0.15				
desertified drylands	40	0.50	1	Woodland	1.2	50
other	60	0.10	1.5	Temp. forest, managed	3.6	140
Pastures	310	0.25				
desertified drylands	180	0.47	5	Woodland	1.2	50
other	130	0.15	4	Temp. forest, managed	3.6	140
Forest lands						
Acid-rain damaged	180	1.00	6	Temp. forest, managed	0	0
Subtotal	*590*	*0.15*	*18*			
Developing countries						
Cropland						
desertified drylands	225	0.50	5	Agroforestry	1	30
Pastures						
desertified drylands	600	0.50	5	Woodland	0.5	15
other pastures	60	0.17	1.5	Mngd. forest/tree crops	5	100
Forest lands						
Clearing f. agriculture	200	0.06				
of which dry woodlands	100	0.03	5	Agroforestry	-0.5	-20
of which other forest land	100	0.03	5	Tree crop agric.	-2.5	-60
Degraded forest areas						
logged over tropic. forests	100	0.50	2	Tropic. forest, managed	5	100
Other degraded areas	50	0.06	1	Industrial plantation	7	80
	50	0.06	1.7	Peri-urban fuelwood lots	7	80
	400	0.48	3.3	Woodland/forests	2	50
Subtotal	*1890*	*0.32*	*29.5*			
Total	**2480**	**0.25**	**47.0**			

[1] Estimate based on typical growth periods and on column 7.
[2] Estimates based on data in Table 3.10, representing net change in carbon storage under steady-state conditions.

annual reforestation or afforestation rate required to accomplish the total plantings within the next 50–100 years.

Overall, some 25 percent of the world's agricultural and forest area would be affected by the scenario efforts. Each year, close to 50 million ha of land would be improved, roughly twice the 20–25 million ha currently clear-cut, degraded, and lost to desertification (see Table 3.12). These afforestation and forest upgrading activities of 50 million ha/yr would largely be in addition to the current worldwide reforestation activity of about 15 million ha/yr (WRI/IIED 1987), which principally maintains a steady-state in affected areas of existing forests. The modest afforestation activity in industrialized countries would increase about tenfold compared to early 1980s levels. Overall, this level of activity represents a very large increase in biospheric maintenance and repair. Much of it would become an integral part of agricultural activities.

Rates of Carbon Fixation from Tree Planting. In relating a tree-planting scenario to the climatic goals outlined above, one must distinguish between the *rate* at which tree planting can fix atmospheric carbon, and the *cumulative carbon storage* achieved per unit of land area once trees have matured and forest cover is maintained at a steady-state level.

The basic calculation in determining the rate of carbon releases or sequestering is to multiply the areas planted by the average rate of fixing achieved on the type of land planted.[28] However, in calculating the combined effect of tree planting done at different times, one must consider the logistics of tree planting and growth. Carbon fixing from tree growth will largely cease after 20–40 years, as trees mature and a steady state is reached.[29]

Forest growth is measured in terms of net primary productivity of plant matter (tons of biomass/ha-year). The carbon portion of this biomass is about 50 percent. Including soil storage, a 60 percent share is a good ballpark figure. In managed temperate forests, 6 t biomass/ha-yr (3.6 t carbon/ha-yr) is a typical value.

Much larger rates of carbon fixing can be achieved with intensive plantation management and with selected species, such as fast-growing eucalyptus or sycamore trees. A species of American sycamore achieves 7.5 t C/ha-yr (Dyson and Marland 1979). In the tropics, the leucaena tree is a similar option. Dyson (1977) proposed a biogenic "carbon bank" (similar to a blood bank) that would store the carbon produced from fast-growing trees and water plants in the form of timber, humus, or peat.

As a supplement to reforestation, the intensive cultivation of water plants has also been proposed (Dyson and Marland 1979). The water hyacinth, a widely distributed freshwater plant originating from South America, can, for instance, store some 6000 t C/km^2/yr, which is about eight times as much as the American sycamore. To sequester the

[28]A second increment of reduction comes from land that would normally have been clear-cut but was preserved.

[29]In adding up contributions to the total rate of carbon fixing, only areas with overlapping growth periods are additive. The maximum rate of carbon fixing as shown in Table 3.13 is calculated on the basis of a 20-year overlap in the growing periods between all areas planted.

roughly 5 billion t of fossil carbon released each year, one would need an area of about 800,000 km^2 for the hyacinths. The total alluvial land in the tropics suitable for hyacinth growth amounts currently to about 2.5 million km^2. However, if this plant were to be used as a carbon sink, one would have to store it after harvest in a manner that avoids releases of methane from its decomposition.

For most afforestation opportunities, carbon fixing based on such high single-species values and exotic schemes is not realistic. For one, they ignore the ecological lesson that intensive monocultures deplete soils and are susceptible to disastrous pest attacks, which have occurred in a number of single-species reforestation projects (Postel and Heise 1988). Leucaena trees can invade land and become a weed problem. The dense growth of hyacinths kills fish and clogs waterways.

A sustainable approach to reforestation would seek to mimic the sophisticated division of labor developed over millennia in natural forests, including the combination of upper-story, middle-story, lower-story, and edge species, different crown sharps to reduce storm damage, and different root systems.

Worldwide, several thousand tree species are available for use in reforestation under the most severe conditions. The efficient and sustainable use of this inherited diversity in a variety of settings still awaits comprehensive research and exploration. Nitrogen-fixing trees can be used to regenerate soils. Trees such as the Nypa palm of the Southeast Asian mangrove forests thrive in areas flooded by saltwater tides, and hold promise for reforestation on salinated soils. They also produce fruit suitable for high-yield alcohol production, as mentioned above.

Furthermore, single-species plantations alter the natural environment in a way that deprives the rural Third World poor of a significant portion of their "gross natural product," which is as important to them as the gross national product is to the cash economy. In many areas of the Third World, the rural poor rely on forests as their apothecary (barks, leaves, and other medicinally useful plant material), food store (nuts, tree fruit, oils, and edible and palatable tropical leaves), utility (fuelwood from fallen branches, and twigs), ranch (animal fodder from leaves, shoots, etc.), and recreational and spiritual centers.

Most of these values are degraded or totally eliminated in a plantation. For example, development-assistance-financed eucalyptus wood-lots in India were based on species that provided no forage for the animals of the poor, as natural forests did. In Africa, women using eucalyptus as cooking fuel have complained that their meals took on a flavor of cough medicine. Understandably, the widespread use of eucalyptus — currently more than 40 percent of all hardwood planting in the tropics—has sparked rural protest movements in a number of countries.[30]

In some circumstances, monoculture plantations may have their uses, at least as a short-term measure. For example, they could be used in commercial woodlots surrounding large cities and for supplying industrial paper, pulp, and timber needs. This would provide relief from deforestation associated with these demands.

What levels of cumulative carbon storage could be achieved through afforestation and forest upgrading in the long term are indicated by the storage values for various land-use types in Table 3.10. At present, net increases of carbon storage per hectare can only be roughly estimated (Houghton et al. 1983).

As a rule, the average carbon storage in a frequent-harvest fuelwood or industrial plantation just before harvest will be lower than in a less intensively managed forest, where in turn it is lower than in a mature natural stand. The qualitative relationships are shown in Figure 3.4. In the figure, the average carbon storage over the harvest and growth cycle of a sustained-yield, short-rotation forest is about half the value achieved just before cutting. Of course, this fraction could be higher under different tree-usage schemes, such as coppicing trees for fuelwood rather than felling them.

Table 3.13 shows the assumptions on carbon-fixing rates and levels used in our scenario estimate.[31]

Reforestation in the Industrialized Countries

Croplands. If beef consumption were to adjust to more moderate levels, and some of the more marginal croplands were set free under a reorganization of the farm subsidy systems,[32] it is estimated that about 100 million ha of cropland (about 20 percent of the OECD cropland or 15 percent of the cropland in the developed world) could be converted to managed temperate forests and woodlands.

---

[30]See, for example, *Development Forum* (July/Aug. 1987).

[31]The carbon-fixing figures include both vegetation and soil storage. The assumed biomass storage in agroforestry is equivalent to 25 percent tree cover. For tree-crop farming, we assume an intermediate value between tropical cropland and primary forests. The carbon-fixing rates in Table 3.13 are estimated from the cumulative carbon storage per hectare shown in the table, assuming typical growth periods for the different kinds of plantings. All figures should be treated as rough estimates and are used here solely to establish an order of magnitude of potential reforestation impacts.

[32]An in-depth discussion of agricultural reform policies that simultaneously addresses ecological, economic, and social issues can be found in Bechmann (1987), who discusses the situation in the FRG and the European Community.

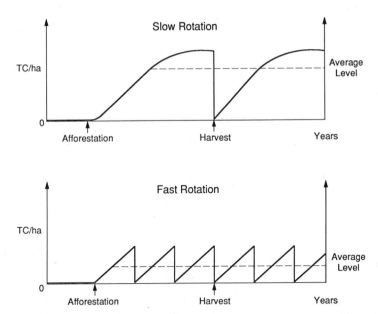

Figure 3.4 *Simplified relationship between harvesting and stored carbon on afforested land used for sustainable-yield forestry.*

Some 240 million ha of industrial-country croplands are in dryland regions, and about 80 million ha are desertified. With proper targeting, the retirement of 40 million ha from dryland croplands could also help fight desertification.

Increases in agricultural productivity brought on by technical innovations would work in the same direction, but have not been included in this estimate. As already noted, the potential for such increases could be large in the socialist countries. However, conventional means of increasing agricultural productivity based on fertilizers and frequent tillage are themselves implicated in climate change (see Section C).

Pastures. Reduced meat and dairy production will mean less need for pasture areas. We assume that about 300 million ha of pastures (about 25 percent of the developed world total) will be afforested. Again, the selection of areas to be put to trees could be geared, in part, to mitigating desertification.[33]

---

[33]About 1 billion ha of industrial country pasture is in semiarid and arid regions. Six hundred million ha of this dryland area is desertified. Short of rainfall constraints, all of the 180 million ha of pasture afforestation could be geared toward desertified areas. In practice, only a fraction of the desertified areas may be able to support woody biomass stands.

Repairing Acid Rain Damage. A further increment of reforestation is the replanting of acid-rain-damaged forests. Possible carbon releases from forest dieback in these damaged areas were not counted in the biospheric release estimates of Houghton et al. (1987). As a consequence, no net carbon-fixing benefits are assigned to this activity relative to current levels.

IC Total. In total, afforestation and reforestation in the industrialized countries would progress at peak rates of about 18 million ha/yr. This compares to a net reforestation rate of 8.4 million ha/yr in these countries during the 1980s (WRI/IIED 1987).

Reforestation in the Developing Countries

Croplands in Desertified Semiarid Regions. Perhaps the most pressing reforestation action needed is in the desertified regions of semiarid Africa. In Table 3.13 we assume that over time, about half the desertified cropland in the Third World semiarid dryland regions will be improved through agroforestry schemes, and that half of the desertified pastures in the Third World will be put under woodlands again and managed in a sustainable manner.

Cattle Ranches. We also assume that clear-cutting of tropical forests in Latin America for cattle ranching is phased out and partially reversed. According to data from the *World Resources Yearbook* (WRI/IIED 1987), about 60 million ha of pasture were added in the tropics in the last 20 years. We assume that an area equivalent to these pastures will be reconverted to managed tropical forests and tree-crop agriculture schemes. The net effect will be the elimination of a sizable source of biospheric carbon, and an increase in carbon storage above current levels.

Cropland Expansion. For the purpose of our reforestation estimate, we assume that the expansion of agricultural land in the Third World can be limited to 200 million ha, as projected for the year 2000. This corresponds to a 40 percent increase in Third World cropland. Losses of arable land are assumed to be compensated for by redirecting some export crop plantation lands toward domestic consumption. The gap between this 40 percent increase and the expected doubling of Third World population is to be closed by productivity increases.

The expansion by 200 million ha is treated as follows: continuing present trends into the near term, some 100 million ha of closed, open, and woodland forest areas are cleared for permanent cropland, and a further 100 million ha is converted to shifting cultivation. Over time, both the existing and these added shifting cultivation areas are converted to tree-crop agriculture and agroforestry schemes. As a result, the net annual losses of forests from the expansion and intensification of shifting cultivation (see Table 3.11) would wane, eliminating that source of net carbon releases. All longer-term food production needs arising in the years beyond 2000 would be met by productivity increases on existing land.[34]

Industrial Wood and Timber Requirements. About 40 percent of industrially harvested wood worldwide is for pulp and paper; the rest is for hardwood and other wood products (Houghton et al. 1983). In tropical hardwood consumption for timber and furniture, the harvested carbon is not released for decades. Here, it is not so much the utilization of the wood, but the destructive impact of harvesting of select species that is worrisome. The curtailment of tropical hardwood imports by industrial countries would help contain industrial logging as Third World pulp and paper needs grow. Recycling of paper products, improved harvesting, and utilization of a broader array of tree species could also greatly relieve these pressures.

We assume that current and future needs for pulp and paper and for industrial hardwood will eventually be met on a renewable basis in 50 million ha of industrial plantations (Postel and Heise 1988). These plantations would be located on already degraded forestland. As a result, the current net carbon release to the atmosphere from industrial wood and timber needs would gradually be eliminated. In time, the plantations would add an increment of carbon storage.

Degraded Tropical Rain Forests. A further element of our scenario is the rehabilitation of about half of the estimated 200 million ha of tropical forests that have been severely degraded, but are still carried as natural forest on the books of FAO statistics. This rehabilitation would increase the carbon storage in these areas above present levels.

Degraded Forests and Shrub Land. We also assume that over time, about half of the 800 million acres of other degraded forests and shrub lands outside agricultural areas will be improved through woodland planting and reforestation.

---

[34]The assumption of a very limited increase in cropland 100 years from now relative to today could include a transient excursion of cropland acreage, followed by reforestation once productivity increases have been more fully realized.

Fuelwood Requirements. According to World Bank estimates, some 55 million ha of fuelwood planting will be needed between now and the turn of the century. Fuelwood project planting rates would have to be increased by a factor of 5 above current levels of about 0.5 million ha/yr to reach this goal. The figure already includes an increase in the efficiency of fuelwood use by 25 percent (Postel and Heise 1988). In the longer term, an even greater need for biomass fuels can be expected.

We assume that 50 million ha will be planted in peri-urban plantations to meet the needs of the landless and of poor city people. All remaining rural needs would be met by the dispersion of agroforestry. In the longer term, gasification schemes using improved bigoas digesters and thermal gasifiers will stabilize biomass fuel requirements and could even provide growing quantities of energy services. In combination, these plantings will eliminate current fuel-related deforestation and, in the longer term, add a net increment of terrestrial carbon storage.

Adding It All Up. Table 3.14 shows in rough numbers what the impact of this scenario would be. Looking at the carbon flux, we see that the above reforestation and afforestation measures would, during the decades of peak planting and regrowth, create a substantial sink of carbon. The maximum rate of carbon sequestering might reach 1.3 btC/yr.[35] If existing carbon pools are stabilized as well, as assumed in the above scenario, it would offset a quarter of the 1985 fossil carbon releases. Two-thirds of this sink would be in developing countries.

In cumulative terms, about 90 billion t of carbon would be sequestered from the atmosphere over the next 50–100 years. This would be about half of the biospheric releases since 1860, about a quarter of total biospheric and fossil releases since 1860, and about 15 percent of the total carbon releases between 1860 and 2100, if the toleration scenario is assumed for fossil fuel consumption.

Conclusions. The above results are based on a global afforestation and reforestation effort that is major by any standard, and would indeed represent a breakthrough in humankind's ability to act effectively as a world community. We can conclude the following:

---

[35]To calculate this maximum value, we assume a 20-year overlap among regrowth periods on all land areas. This means that the contribution from a given land category is 20 times the annually reforested increment. We also treat all land areas as additive, though there might in practice be some overlap.

TABLE 3.14 Sequestering of Atmospheric Carbon through Tree Planting: Rough Upper-Limit Estimate

Reforestation or other land-use conversion on (current land use)	Conversion to (land-use)	Max. rate of C fixing[a]		Sequestered carbon[b]	
		Absolute bt C/yr	Fraction of 1980's fossil C flux	Absolute bt C	Fraction of 1860-1980 biosph. C rel.
Industrialized countries					
Cropland					
desertified drylands	Woodland	0.02	0.00	2	0.01
other	Temp. forest, managed	0.11	0.02	8.4	0.05
Pastures					
desertified drylands	Woodland	0.12	0.02	9	0.05
other	Temp. forest, managed	0.29	0.05	18.2	0.10
Forest lands					
Acid-rain damaged	Temp. forest, managed	0	0.00	0	0.00
Subtotal		*0.54*	*0.10*	*38*	*0.21*
Developing countries					
Cropland					
desertified drylands	Agroforestry	0.10	0.02	6.75	0.04
Pastures					
desertified drylands	Woodland	0.05	0.01	9	0.05
other pastures	Mngd. forest/tree crops	0.15	0.03	6	0.03
Forest lands					
Forest clearing f. agriculture					
dry woodlands	Agroforestry	-0.05	-0.01	-2	-0.01
other forest land	Tree crop agric.	-0.25	-0.05	-6	-0.03
Degraded forest areas					
logged over tropical forests	Tropic. forest, managed	0.20	0.04	10	0.06
Other degraded areas	Industrial plantation	0.14	0.03	4	0.02
	Peri-urban fuelwood lots	0.24	0.04	4	0.02
	Woodland/forests	0.13	0.02	20	0.11
Subtotal		*0.71*	*0.13*	*52*	*0.29*
Total		**1.3**	**0.24**	**89**	**0.50**

[a]Calculated from Table 3.13 assuming a 20-year overlap in peak growth periods. Rate of C fixing = (column 4) × (column 6) × 20.
[b]Calculated from Table 3.13. Sequestered carbon = (column 2) × (column 7).

- A very major and sustained afforestation and reforestation effort would have a substantial impact on atmospheric carbon concentration and could sequester perhaps a quarter of all fossil and biospheric carbon released to the atmosphere to date.
- Such an effort could slow the growth of fossil-related climate risks, though offsets of current fossil CO_2 releases would at best be partial and temporary.
- The goal of restoring preindustrial atmospheric carbon concentrations by means of tree planting is unattainable within the next century. Though a major effort could offset a significant portion of

historic carbon releases, it would still be overwhelmed by the further fossil releases that are unavoidable during a fossil phase-out.

- The much more modest goal of merely rendering the biosphere neutral and maintaining terrestrial carbon pools at approximately current levels is itself an ambitious target and will require unprecedented policies and action.

Thus, even under optimistic assumptions about tree planting worldwide, trees cannot, in the aggregate, substitute for the fossil-displacing investments required under the toleration scenario. Tree planting must be done in addition. This finding suggests that no trade-offs should be made between curtailing fossil carbon releases and biospheric sequestering *unless and until zero* net releases from the biosphere have been achieved and recent levels of global biotic carbon storage have been reestablished.

4. Some Policy Implications

How likely is it that the world will be able, within the next few decades, to more than neutralize current biospheric fluxes? Just neutralizing the current rate of carbon pool destruction of about 20–25 million ha/year (Table 3.12) will require implementation of roughly half the measures listed in our scenario. A look at the social and political obstacles to reforestation provides a further sobering perspective.

The Politics of Tree Planting. The political challenge of implementing even the more modest goal of biosphere neutralization is daunting. Our scenario goes far beyond the modest beginnings of international action, as represented by the Tropical Forest Action Plan (TFAP) sponsored by the World Bank, UNEP, and the U.S. Agency for International Development (WRI 1986).

While an encouraging beginning, this plan provides a one-time sum of $8 billion of capital over five years, or $1.6 billion per year. If current trends are to be slowed and turned around in a timely manner, much larger sums of capital are likely to be needed—on the order of 10–20 billion dollars per years.[36] More broadly, the developing coun-

---

[36]Using a very favorable figure of $1 per tree for the average planting, establishment, and maintenance cost, and a density of 400 trees per ha on average, tree planting on 47 million ha of land each year (Table 3.13) would cost $18.8 billion, or more than 10 billion dollars per year if planting is slowed to continue till 2100. Note, however, that the net cost may be much smaller over time, since the first cost of tree planting will in many settings be offset by significant economic returns (Leach and Mearns 1988).

tries will need greatly stepped-up assistance to develop efficient, environmentally and socially benign technologies in their energy, agriculture, and forestry sectors.

Such change may be slow in coming, for the following reasons:

- The current destruction of the biosphere cannot be seen separately from the pervasive inequities of the current world economic order, which has created huge debt burdens on Third World countries, a widening north–south gap in living standards, and a massive crisis in the basic needs of the developing world. For example, it is the debt crisis that directly drives much of the destruction of tropical forests, which are mined for short-term foreign-exchange earnings.
- Nor is it possible to implement a global afforestation strategy without the active involvement and participation of people and local communities everywhere, notably the rural poor in the developing world. In this regard, efforts like the TFAP may err in the direction of overreliance on multinationally organized, commercialized forestry, which has often proven antithetical to such mobilization.[37]
- A major reason for Third World poverty and environmental destruction is the highly inequitable distribution of land ownership and income in most of these countries. This inequitable and environmentally destructive internal social order is widely maintained by force. Most peoples of the Third World live under undemocratic, repressive regimes that are prone to see grass-roots initiatives of any kind as a potential threat.

In general, there is a danger of overreliance on well-intentioned proposals that are modeled on implementation options in the industrialized countries. Reforestation and afforestation in the developing countries is unlikely to succeed if pursued merely as a single-issue campaign or a commercial activity, without solutions to the broader social questions of basic-needs-oriented, sustainable development. For all practical purposes, augmenting the world's carbon pools beyond present levels will therefore be a long-term prospect.

---

[37]An example of an alternative approach is the proposal by India's Center for Science and the Environment (CSE) to settle 30 million landless families on 2-ha plots of degraded yet recoverable land, and let them produce a livelihood, including cash crop trees for pulp and paper needs of the commercial sector, but in balance with all the other requirements of a sustainable agroforestry system. Such a policy, argues the CSE, would bring reforestation goals in line with the social requirements of sustainable development.

Financing Mechanisms. One way in which the large sums of capital needed for tree planting could be raised is by levying a tree-planting surcharge on nonrenewable energy use in industrialized countries. For example, one could envision that every nonrenewables based new power plant built in an industrialized nation would be taxed with a surcharge to finance reforestation, whether on industrial-country cropland or in the Third World.

One U.S. supplier of cogeneration power plant equipment, Applied Energy Systems, is already voluntarily applying such a (less than 5 percent) price surcharge on their projects for Third World reforestation purposes. In the United States, it also has been proposed to levy a fine on new power plants to finance the reforestation of 16 million ha of surplus cropland set aside under the country's Conservation Reserve Program (Dudek 1988).

A further way of raising funds would be to convert Third World debt payments into contributions to an international climate protection fund that would finance afforestation and reforestation projects in developing countries, as well as transfers of energy-efficient technology. Alternatively, developing countries could be given debt relief in proportion to domestic reforestation efforts. Tree-planting campaigns would require no foreign exchange and could mobilize the Third World's domestic human resources both for development and for global environmental goals.

However, the main problem may not be how to raise the necessary funds, but how to find a way of disbursing them effectively and productively. In addition, problems of monitoring debt-for-forest trades could be significant. Furthermore, if debt-for-forest swaps are promoted with excessive pressure, they might be seen as yet another form of northern interference in Third World affairs. This could stir resentment and backlash, particularly in view of the fact that most of the climate-warming threat has been caused by the industrialized countries (see Chapter 5).

Trade-Off Issues. It is evident from our previous discussions that the toleration scenario, which assumes a neutral biosphere, represents an ambitious target. If investments in fossil carbon abatement were traded off for investments in tree planting, rather than being pursued in tandem, this target could be badly missed.

For instance, under schemes that allow fossil–biotic offsets and apply a surcharge per unit of net carbon emitted, there could be an incentive for the individual utility or industrial power plant investor to engage in tree planting *instead* of efficiency and renewables programs. A possible solution to this problem might be to require more than one unit of biotic carbon storage per unit of fossil carbon offset (see also Chapter 7).

There is no doubt carbon-surcharge/tree-planting trade-off mechanisms could be an effective means of administering policies and of raising capital for reforestation projects. However, one must carefully distinguish between trade-off arrangements that serve as administrative tools for stimulating individual economic actors to participate in a societal long-term strategy, and the overall target setting for policy purposes. Policymakers must still determine what quantities of biospheric and fossil emission reductions should be achieved overall, and over what periods of time, given the goal of climate stabilization.

REFERENCES

Agarwal, A. (1985), *Energy for Sustainable Development, Part Three*, Environment Liaison Centre, Nairobi, Kenya.

Akbari, H., et al. (1988), The Impact of Summer Heat Islands on Cooling Energy Consumption and Global CO_2 Concentrations, LBL-25179, Lawrence Berkeley Laboratory, Berkeley, CA, Apr.

Allen, A. Y. (1968), "Maize Diamonds," *Kenya Farmer*, January (Nairobi).

Andrae, G., and B. Beckman (1985), *The Wheat Trap: Bread and Underdevelopment in Nigeria*, with Scandinavian Institute of African Studies, Zed Books, London.

Bach, W. (1984, *Our Threatened Climate*, Reidel, Dordrecht, the Netherlands.

Bach, W. (1985), "Forest Dieback: Extent of Damages and Control Strategies," *Experientia*, Vol. 41, pp. 1095–1104.

Banerjee, S., and S. Kothari (1985), "Food and Hunger in India," *Ecologist*, Vol. 15, No. 5/6, pp. 257–260.

Bechmann, A. (1987), *Landbau-Wende: Gesunde Landwirtschaft — Gesunde Emaehrung*, S. Fischer Verlag, Frankfurt.

Billings, W. D., et al. (1982), "Arctic Tundra: A Source or Sink for Atmospheric Carbon Dioxide in a Changing Environment?" *Oecologia*, Vol. 53, pp. 7–11.

Billings, W. D., et al. (1983), "Increasing Atmospheric Carbon Dioxide: Possible Effects on Arctic Tundra," *Oecologia*, Vol. 58, pp. 286–289.

Billings, W. D., et al. (1984), "Interaction of Increasing Atmospheric Carbon Dioxide and Soil Nitrogen on the Carbon Balance of Tundra Microcosms," *Oecologia*, Vol. 65, pp. 26–29.

Bingemer, J., and P. J. Crutzen (1987), "The Production of Methane from Solid Waste," *Journal of Geophysical Research*, Vol. 92, pp. 2181–2187.

Bossel, H., et al. (1986), *Technologiefolgeabschaetzungen fur die Landwirtschaftliche Produktion*, Institut für Systemanalyse und Prognose (ISP), Hanover, FRG.

Bremner, J. M., and A. M. Blackmer (1978), "Nitrous Oxide: Emission from Soils during Nitrification of Fertilizer Nitrogen," *Science*, Vol. 199, pp. 295–296.

Brody, J. E. (1985), "Organic Farming Moves into the Mainstream," *New York Times* (Oct. 8), p. 20.

Brown, S., and A. E. Lugo (1982), "The Storage and Production of Organic Matter in Tropical Forests and Their Role inthe Global Carbon Cycle," *Biotropica*, Vol. 14, pp. 161–187.

Brown, S., and A. E. Jugo (1984), "Biomass of Tropical Forests: A New Estimate Based on Volumes," *Science*, Vol. 223, pp. 1290–1293.

Bull, D. (1982), *A Growing Problem: Pesticide Use and the Third World Poor*, OXFAM, Oxford, Great Britain.

Cicerone, R., and J. D. Shetter (1981), "Sources of Atmospheric Methane: Measurements in Rice Paddies and a Discussion," *Journal of Geophysical Research*, Vol. 86, pp. 7203–7209.

CMA (1987 a,b), "Scientific Information on Stratospheric Ozone Change," Chemical Manufacturers' Association, paper presented to UNEP.

Crutzen, P. J. (1983), "Atmospheric Interactions—Homogeneous Gas Reactions of C, N, and S Containing Compounds," in B. Bolin and R. B. Cook, eds., *The Major Biochemical Cycles and Their Interactions*, Wiley, New York, pp. 67–112.

Crutzen, P. J. (1986), *Globale Aspekte der atmosphärischen Chemie. Natürliche und anthropogene Einflüsse*, Vorträge No. 347, Rheinisch-Westfälische Akademie der Wissenschaften, Westdeutscher Verlag, Opladen, Germany, pp. 41–72.

Crutzen, P. J. (1987), "Role of Tropics in Atmospheric Chemistry," in R. E. Dickinson, ed., *The Geophysiology of Amazonia: Vegetation and Climate Interactions*, Wiley, New York, pp. 107–131.

Crutzen, P. J., and T. E. Graedel (1986), "The Role of Atmospheric Chemistry in Environment-Development Interactions," in W. C. Clark and R. E. Munn, eds., *Sustainable Development of the Biosphere*, Cambridge University Press, New York, pp. 213–250.

Crutzen, P. J., et al. (1979), "Biomass Burning as a Source of Atmospheric Gases CO, H_2, N_2O, NO, CH_3Cl and COS," *Nature*, Vol. 282, pp. 253–256.

Crutzen, P. J., et al. (1986), "Methane Production by Domestic Animals, Wild Ruminants, Other Herbivorous Fauna and Humans," *Tellus*, Vol. 38B, pp. 271–184.

DECHEMA, (1987), *Anthropogene Beeinflussung der Ozonschicht. 6. DECHEMA-Fachgespräche Umweltschutz*, 16.–17.12.1987, Frankfurt.

Drenham, B., and C. Hines (1984), *Agribusiness in Africa*, Earth Resources Research Ltd., London.

Dudek, D. (1988),"Offsetting New CO_2 Emissions," Environmental Defense Fund, Washington, D.C., Sept., unpublished paper.

Dyson, F. J. (1977), "Can We Control the Carbon Dioxide in the Atmosphere?" *Energy*, Vol. 2, pp. 287–291.

Dyson, F. J., and G. Marland (1979), "Technical Fixes for the Climatic Effects of CO_2," in W. P. Elliott and L. Machta, eds., *Workshop on the Global Effects of Carbon Dioxide from Fossil Fuels*, CONF-770385-001, U.S. Department of Energy, Washington, D.C., pp. 111–118.

Edmonds, J. A., and G. Marland (1986), *The Energy Connection to Climate Change: Gaseous Emissions*, Institute for Energy Analysis, Oak Ridge, TN.

Egger, K. (1978), "Ecofarming—Entwicklungsstrategie für Problemgebiete?" *Entwicklung und Ländlicher Raum*, Vol. 2, pp. 10–14.

Ehhalt, D. H. (1985), "Methane in the Global Atmosphere," *Environment*, Vol. 27, No. 10, pp. 6–12, 30–33.

Enquête-Kommission (1988), *Erster Zwischenbericht, Enquete-Kommission Vorsorge zum Schutz der Erdatmosphaere*, Deutscher Bundestag, Bonn, Nov. 2.

EPA (1988), *Future Concentrations of Stratospheric Chlorine and Bromine*, EPA Report 400/1-88/005, U.S. Environmental Protection Agency, Washington, D.C., August.

Fabian, P. (1988), *Stellungnahme zum Fragenkatalog Fluorchlorkohlenwasserstoffe und Stratosphärisches Ozon*, Kommissionsdrucksache, Bonn, Nov. 8.

FAO (1978), *The State of Food and Agriculture 1977*, United Nations Food and Agriculture Organization, Rome.

FAO (1980), *Soil and Water Conservation*, United Nations Food and Agriculture Organization, Committee on Agriculture, Rome, Nov.

FAO/UNEP (1981), *Tropical Forest Resource Assessment*, United Nations Food and Agriculture Organization, Rome.

Fraser, P. J., et al. (1986), "Termites and Global Methane—Another Assessment," *Journal of Atmospheric Chemistry*, Vol. 4, pp. 295–310.

George, S. (1976), *How the Other Half Dies*, New York.

Giesen, K. (1986), "Stand der Waldschäden und der Gegenmaßnahmen in den Europäischen Ländern," *Allgemeine Forstzeitschrift*, Vol. 41, pp. 1006–1007.

Global 2000 (1981), *Der Bericht an den Präsidenten*, 6. Aufl. Zweitausend, Frankfurt.

Goldsmith, E. (1985),"Is Development the Solution or the Problem?" *Ecologist*, Vol. 15, No. 5/6, pp. 210–219.

Groffman, P., and P. Hendrix (1987), "Nitrogen Dynamics in Conventional and No-Tillage Agroecosystems with Inorganic Fertilizer or Legume Nitrogen Inputs," *Plant and Soil*, Vol. 97, pp. 315–322.

Hammitt, J. K., et al. (1986), *Product Uses and Market Trends for Potential Ozone-Depleting Substances: 1985–2000*, The Rand Corporation, Santa Monica, CA.

Hammitt, J. K., et al. (1987), "Future Emission Scenarios for Chemicals That May Deplete Stratospheric Ozone," *Nature*, Vol. 330, pp. 711–716.

Hampicke, U., and W. Bach (1980), "Die Rolle Terrestrischer Okosysteme im Globalen Kohlenstoffkreislauf," *Münstersche Geographische Arbeiten*, Vol. 6.

Hoffman, J., and M. J. Gibbs (1988), *Future Concentrations of Stratospheric Chlorine and Bromine*, EPA 400/1-88/005, U.S. Environmental Protection Agency, Office of Air and Radiation, Washington, D.C.

Houghton, R. A. (1986), "Estimating Changes in the Carbon Content of Terrestrial Ecosystems from Historical Data" in J. R. Trabalka and D. E. Reichle, eds., *The Changing Carbon Cycle: A Global Analysis*, Springer, New York.

Houghton, R. A., J. E. Hobbie, and J. M. Melillo (1983), "Changes in the Carbon Content of Soils between 1860 and 1980: A Net Release of CO_2 to the Atmosphere," *Ecological Monographs*, Vol. 53, No. 3, pp. 235–262.

Houghton, R. A., et al. (1985), "Carbon Dioxide Exchange between the Atmosphere and Terrestrial Ecosystems," in J. R. Trabalka and D. E. Reichle, eds., *The Changing Carbon Cycle: A Global Analysis*, Springer, New York.

Houghton, R. A., et al. (1987), "The Flux of Carbon from Terrestrial Ecosystems to the Atmosphere in 1980 Due to Changes in Land Use: Geographic Distribution of the Global Flux," *Tellus*, Vol. 39B, No. 1–2, pp. 122–139.

IPCC (1990), *Greenhouse Gases and Aerosols*, Section 1, 2nd draft report of Working Group 1, International Panel on Climate Change, Geneva, Mar. 9.

Jackson, W. (1980), *New Roots for Agriculture*, Friends of the Earth, San Francisco, and Land Institute, Salinas, Kansas.

Kavanaugh, M. (1987), "Estimates of Future CO, N_2O, and NO_x Emissions from Energy Combustion," *Atmospheric Environment*, Vol. 21, No. 3, pp. 463–468.

Khalil, M. A. K., and R. A. Rasmussen (1983), "Sources, Sinks and Seasonal Cycles of Atmospheric Methane," *Journal of Geophysical Research*, Vol. 88, pp. 5131–5144.

Khalil, M. A. K., and R. A. Rasmussen (1985), "Causes of Increasing Atmospheric Methane: Depletion of Hydroxyl Radicals and the Rise of Emissions," *Atmospheric Environment*, Vol. 19, pp. 397–407.

Krause, F., and J. Koomey (1989), "Unit Cost of Carbon Savings from Urban Tree Planting, Rural Tree Planting, and Energy Efficiency Improvements," in *Saving Energy and Reducing Atmospheric Pollution by Control-*

ling Summer Heat Islands, Proceedings of a workshop held at Lawrence Berkeley Laboratory, Berkeley, CA, Feb. 23–24.

Kuhler, M., J. Kraft, H. Klingenberg, and D. Schürmann (1985), "Natürliche und anthropogene Emissionen," *Automobil Industrie*, Vol. 30, No. 2, pp. 1–12.

Lashof, D. A. (1989), "The Dynamic Greenhouse: Feedback Processes That May Influence Future Concentrations of Atmospheric Trace Gases and Climate Change," *Climatic Change*, Vol. 14, pp. 213–242, Revised draft, Jan. 1989.

Lashof, D. A., and D. A. Tirpak, eds. (1989), *Policy Options for Stabilizing Global Climate*, Draft report to the U.S. Congress, U.S. Environment Protection Agency, Office of Policy, Planning, and Evaluation, Feb.

Leach, G. (1982), "Cross-Country Study: Energy for Agriculture," Paper presented at the Princeton Workshop on End-Use Focused Energy Strategies, Princeton, NJ, Apr. 21–28.

Leach, G., and R. Mearns (1988), *Beyond the Woodfuel Crisis: People, Land, and Trees in Africa*, Earthscan, London.

Logan, J. A. (1983), "Nitrogen Oxides in the Troposphere: Global and Regional Budgets," *Journal of Geophysical Research*, Vol. 80, pp. 10,785–10,807.

Logan, J. A., et al. (1981), "Tropospheric Chemistry: A Global Perspective," *Journal of Geophysical Research*, Vol. 86, pp. 7210–7254.

Lovins, A. B., L. H. Lovins, F. Krause, and W. Bach (1981), *Least-Cost Energy: Solving the CO_2-Problems*, Brick House, Andover, MA. (German version: *Wirtschaftlichster Energieeinsatz: Lösung des CO_2 Problems*, Karlsruhe, FRG, 1983.)

MacDonald, G. J. (1985), *Climate Change and Acid Rain*, The MITRE Corporation, McLean, VA, Nov.

Makhijani, A., Makhijani, A., and A. Bickel (1988), *Saving Our Skins: Technical Potential and Policies for the Elimination of Ozone-Depleting Chlorine Compounds*, Environmental Policy Institute and Institute for Energy and Environmental Research, Washington, D.C., Sept.

Marland, G. (1985), *Anthropogenic Impacts on the Atmospheric Concentration of Nitrous Oxide*, Institute for Energy Analysis, Oak Ridge, TN.

Marland, G., et al. (1988), *Estimates of CO_2 Emissions from Fossil Fuel Combustion and Cement Manufacturing Using the United Nations Energy Statistics and the U.S. Bureau of Mines Cement Manufacturing Data*, ORNL-3176, Oak Ridge National Laboratory, Oak Ridge, TN, Sept.

Mathews, E. (1983), "Global Vegetation and Land Use," *Journal of Climate and Applied Meteorology*, Vol. 22, pp. 474–487.

Mathews, E., and I. Fung (1987), "Methane Emissions from Natural Wetlands: Global Distribution, Area and Environmental Characteristics of Sources," *Global Biogeochemical Cycles*, Vol. 1, pp. 61–86.

McElroy, M. B., and S. C. Wofsy (1987), "Tropical Forests: Interactions with the Atmosphere," in G. T. Prance, ed., *Tropical Forests and World Atmosphere*, Symposium Volume, Westview Press, Boulder, CO.

McFarland, R. (1988), Testimony before the FRG Enquête Kommission, Bonn, Chemical Manufacturers Association.

Miller, A. S., and I. M. Mintzer (1986), *The Sky is the Limit: Strategies for Protecting the Ozone Layer*, Research Report No. 3, World Resources Institute, Washington, D.C.

Moore-Lappé, F. and J. Collins (1979), *Food First*, New York.

Moore-Lappé, F., J. Collins, and D. Kinley (1981), *Aid as Obstacle*, New York.

Mueller-Elze, R., and W. Bach (1986), *Gesunder Landbau—gesunde Ernährung," Fischer Verlag* 4099, Frankfurt.

Myers, N. (1981), "The Hamburger Connection: How Central America's Forests Become North America's Hamburgers," *Ambio*, Vol. 10, pp. 3–8.

NAS (1974), *More Water for Arid Lands: Promising Technologies and Research Opportunities*, National Academy of Sciences, Washington, D.C.

NAS (1984), *Global Tropospheric Chemistry: A Plan For Action*, National Academy of Sciences, Washington, D.C.

Olson, J. S., et al. (1985), "The Natural Carbon Cycle," in J. R. Trabalka, ed., *Atmospheric Carbon Dioxide and the Global Carbon Cycle*, DOE/ER/0239, U.S. Department of Energy, Dec.

Pilarski, M., ed. (1988), *International Green Front Report*, Friends of the Trees, Chelan, Washington.

Postel, S., and L. Heise (1988), *Reforesting the Earth*, Worldwatch Paper 83, Apr.

Ramanathan, V., et al. (1985), "Trace Gas Trends and Their Potential Role in Climate Change," *Journal of Geophysical Research*, Vol. 90, pp. 5547–5566.

Redclift, M. (1984), *Development and the Environmental Crisis*, New York.

Rejinders, L. (1987), "Data on Chlorofluorocarbons," Utrecht, unpublished report.

Rotty, R. M., and C. D. Masters (1985), "Carbon Dioxide from Fossil Fuel Combustion: Trends, Resources, and Technological Implications," in J. R. Trabalka, ed., *Atmospheric Carbon Dioxide and the Global Carbon Cycle*, DOE/ER/0239, U.S. Department of Energy, Washington, D.C., Dec.

Rowland, F. S. (1988), *Stellungnahme zum Fragenkatalog Fluorchlorkohlenwasserstoffe und Stratosphärisches Ozon*, Kommissionsdrucksache, Bonn, Sept. 11.

Salati, E. (1987), Information presented at the International Workshop on Climate Variability and Food Security in Developing Countries, New Delhi, Fundacao Salim Farah Maluf, São Paulo, Feb. 6–9, 1987.

Sedego, M. (1981), "Contribution a la Valorisation des Residus Culturaux en Sol Ferrugineux et Sous Climat Semi-Aride," Doctoral thesis, University of

Nancy, France. Quoted in H. W. Ohm and J. G. Nagy, eds., *Appropriate Technologies or Farmers in Semi-Arid West Africa*, Purdue University, International Program in Agriculture, West Lafayette, Indiana.

Seiler, W. (1984), "Contributions of Biological Processes to the Global Budget of CH_4 in the Atmosphere," in M. J. Klug and C. A. Reedy, eds., *Current Perspectives in Microbial Ecology*, American Society for Microbiology, Washington, D.C.

Seiler, W. (1985), "Cycles of Radiatively Important Trace Gases (CH_4, N_2O)," in *The Impact of an Increased Atmospheric Concentration of Carbon Dioxide on the Environment*, WMO/ICSU/UNEP, Geneva.

Seiler, W., et al. (1984), "Methane Emission from Rice Paddies," *Journal of Atmospheric Chemistry*, Vol. 1, pp. 241–265.

Singh, H. B. (1987), "Reactive Nitrogen in the Troposphere," *Environmental Science and Technology*, Vol. 21, pp. 320–327.

Slemr, F., and W. Seiler (1984), "Field Measurements of NO and NO_2 Emissions from Fertilized and Unfertilized Soils," *Journal of Atmospheric Chemistry*, Vol. 2, pp. 1–24.

Toepfer, K. (1988), *Referat vor der Enquete Kommission*, Vorsorge zum Schutze der Erdatmosphare, Bonn.

Trabalka, J. R., ed. (1985), *Atmospheric Carbon Dioxide and the Global Carbon Cycle*, DOE/ER/0239, U. S. Department of Energy, Washington, D.C., Dec.

Trabalka, J. R., and D. E. Reichle (1986), *The Changing Carbon Cycle: A Global Analysis*, Springer, New York.

Trabalka, J. R., et al. (1985), "Human Alterations of the Global Carbon Cycle and the Projected Future," in J. R. Trabalka, ed., *Atmospheric Carbon Dioxide and the Global Carbon Cycle*. DOE/ER/0239, U.S. Department of Energy, Washington, D.C., Dec.

UNEP (1987), *Environmental Data Report*, United Nations Environment Program, New York/Nairobi.

Unnasch, S., C. B. Moyer, D. D. Lowell, and M. D. Jackson (1989), *Comparing the Impact of Different Transportation Fuels on the Greenhouse Effect*, California Energy Commission, Sacramento, Mar.

Wade, E. (1981), "Fertile Cities," *Development Forum*, Sept./Dec.

Weiss, R. S. (1981), "The Temporal and Spatial Distribution of Tropospheric Nitrous Oxide," *Journal of Geophysical Research*, Vol. 86, pp. 7185–7195.

Winstanley, D. (1983), "Desertification: A Climatological Perspective," in *Origin and Evolution of Deserts*, Chapter 7.

WMO (1985), *Atmospheric Ozone 1985*, 3 Vols. Global Ozone Research and Monitoring Project, Report No. 16, World Meteorological Organization, Geneva.

Wolf, Edward C. (1986), "Byond the Green Revolution: New Approaches for

Third World Agriculture," Worldwatch Paper 73, Worldwatch Institute, Washington, D.C., Oct.

Woodwell, G. M. (1983), "Biotic Effects on the Concentration of Atmospheric Carbon Dioxide: A Review and Projection," in *Changing Climate*, National Academy of Sciences, Washington, D.C., pp. 216–241.

Woodwell, G. M. (1986), "Changes in the Area of Forests in Rondonia, Amazon Basin, Measured by Satellite Imagery," in *The Changing Carbon Cycle: A Global Analysis*, J. R. Trabalka and D. E. Reichle, eds., Springer, New York.

Woodwell, G. M. (1987), "Forests and Climate: Surprises in Store," *Oceanus*, Vol. 29, No. 4 (Winter 1987), pp. 70–75.

WRI (1986), *Tropical Forests: A Call for Action*, World Resources Institute, Washington, D.C.

WRI/IIED (1987), *World Resources Yearbook 1987*, World Resources Institute, Washington, D.C., and International Institute for Environment and Development, London.

WSJ (1989), "Big Firms get High on Organic Farming," *Wall Street Journal*, March 21, p. B1, and "Back to the Future: A Movement to Farm without Chemicals Makes Surprising Gains," May 11, p. A1.

Wuebbles, D. J., and J. Edmonds (1988), "A Primer on Greenhouse Gases," U.S. Department of Energy, Washington, D.C.

4

How Much Fossil Fuel Can Still Be Burned?

A. INTRODUCTION

In this chapter, we use the scenario calculations in Chapter 2 and the discussion of nonfossil trace gases in Chapter 3 to determine an approximate *concentration limit for carbon dioxide.* This concentration limit is then translated into a global *fossil carbon budget.* The next step is to calculate how much *fossil-derived energy* this budget represents, assuming different mixes of coal, oil, and gas. Finally, we explore what share of current global fossil resources can still be burned under this carbon budget.

The central theme in this investigation is the conversion of a carbon dioxide concentration limit into an emission budget. For many purposes, the relationship between carbon dioxide concentrations and fossil fuel consumption can be approximated by simple conversion factors for the energy content of fossil fuels, their carbon content, and the fraction of fossil carbon dioxide releases that are retained in the atmosphere (the airborne fraction, or AF). These simplified calculations are presented in Section C.

A more sophisticated treatment of this conversion involves models of the global carbon cycle. Among currently used dynamic models, which differ greatly in complexity, there is considerable divergence in calculated carbon dioxide concentrations for a given emission increment. This divergence, in turn, varies with the emission path. In Chapter 2, we had used only one model that is considered one of the best in terms of reproducing available empirical data. A sensitivity analysis based on consideration of other models is now in order.

Similarly, the analysis in Chapter 2 considered only three published energy scenarios. A number of other global energy trajectories have been important in the international discussion, and are now incorporated into our sensitivity analysis. In Chapter 2, we also did not present variations in the timing of a fossil fuel phaseout under the toleration scenario. Such fossil phaseout variations are, however, of great significance in deriving fossil carbon budgets and reduction milestones under a tight warming limit. They are also important in testing a key approximation made in the WERM (warming limit/emission budget/reduction milestone method) framework, that is, the use of one and the same carbon budget in devising alternative phaseout schedules.

B. CONCENTRATION LIMIT FOR CARBON DIOXIDE

1. Changing Meaning of Carbon Dioxide Doubling

Early discussions of the greenhouse effect mainly focused on risks from carbon dioxide. They led to a consensus that a doubling of carbon dioxide concentration to 550–580 ppm represented the approximate threshold for definite climate deterioration. It is useful to relate our warming limit analysis to this reference point, for even the conventional doubling criterion has undergone major reinterpretation.

It was demonstrated in Chapter 2 that when other trace gases are taken into account, the date by which equivalent carbon dioxide concentrations double is advanced dramatically. To see this compression in concentration terms, it is useful to reinspect Table 2.4, which shows both CO_2 and equivalent CO_2 concentrations. The influence of non-CO_2 greenhouse gases is measured by the difference between these two numbers. For the medium scenarios C and D, the concentration of CO_2 alone is 450 ppm in 2030. The other greenhouse gases add anywhere from 107 to 179 ppm, for an equivalent total of 557–629 ppm.

If the threshold risk had originally been defined by a doubling, that is, 580 ppm, the effective doubling threshold for CO_2 alone would decrease to the range of $(580 - 179) = 401$ ppm to $(580 - 107) = 473$ ppm. If we use the lower value for the range of estimates for preindustrial CO_2 concentrations, that is, 275 ppm, the doubling level would be 550 ppm, and the equivalent concentration level for CO_2 becomes about 370–440 ppm. Thus, it can be plausibly argued that

- Under conventional projections for nonfossil trace-gas emissions, risk levels associated in the past with a doubling of CO_2 concentration may now be associated with a CO_2 concentration as low as 370–470 ppm.

But more important than the doubling standard and its equivalents is the question of what CO_2 concentration is compatible with the warming limit and toleration scenario defined in Chapters 1 and 2.

2. CO₂ Concentration Limit under the Toleration Scenario

As discussed in Chapter 2, the warming limit translates into a peak radiative forcing limit equivalent to about 430–450 ppm of CO_2 under risk-minimizing assumptions about climate sensitivity. Furthermore, the discussion in Chapters 2 and 3 shows that trace gases other than CO_2 from fossil fuel combustion will contribute at least 50–70 ppm (under scenario E and F) of equivalent CO_2 even if far-reaching control measures are implemented. The lower value is based on a phaseout of CFCs by around the turn of the century. Thus, we can conclude that

- Under a global toleration scenario, the upper limit for atmospheric CO_2 concentrations due to fossil carbon releases is about 380–400 ppm.
- This upper limit applies under the assumption that contributions to radiative forcing from other trace gases will be limited to no more than 50–70 ppm equivalent CO_2.

To the degree that control of other trace gases is not as fully achieved as stated, the concentration limit for fossil carbon dioxide will need to be reduced, or the risk ceiling for climate change as assumed in this report will be exceeded.

C. FOSSIL RESERVES, ENERGY CONTENT, CARBON CONTENT, AND AIRBORNE FRACTION

1. Energy and Carbon Content of Fossil Fuels

An important feature of fossil fuels is that the amount of carbon released per unit of energy obtained (the carbon burden of the fuel)

TABLE 4.1 Energy, Tons of Carbon, and ppm of Carbon Dioxide [1]-[6]

	ppmv	billion t C	TW-y coal	TW-y oil	TW-y gas
ppmv	1	3.85	5.056	6.458	8.960
billion t C	0.26	1	1.313	1.677	2.327
TW-y coal	0.198	0.761	1	1.277	1.772
TW-y oil	0.155	0.596	0.783	1	1.388
TW-y gas	0.112	0.430	0.564	0.721	1

[1] Assuming 55% atmospheric retention. See Section C for a discussion of the applicability of this conversion factor.
[2] Carbon factors from Marland and Rotty (1984), adjusted for complete combustion.
[3] Each ton of carbon dioxide counted as carbon only.
[4] In conversion among energy units, TW-yr stands for carbon content of 1 Terawatt-year.
[5] Energy units based on higher heating value (HHV).
[6] ppm figures are volume-based.

differs widely.[1] These differences are shown in the form of a simple matrix in Table 4.1. To read this matrix, we may start, for example, with natural gas in the lowest row and last column. Here, the carbon content of 1 Terawatt-year (TW-yr) of natural gas is given the index 1.0. Moving upward in the last column, the next row gives the carbon content of 1 TW-yr of oil relative to 1 TW-yr of natural gas. The index is 39 percent higher. Going up another row, we see that the carbon burden of coal is 77 percent higher.

Put another way, burning only gas would produce 77 percent more energy than burning only coal for the same carbon budget. At 39 percent above the burden for gas, oil falls almost exactly between coal and gas.

Carbon Content of Fossil Fuel Reserves and Resources. Table 4.2 shows estimates by Haefele et al. (1981) of global fossil resources by fuel and recovery cost. The energy unit shown is the number of

---

[1] The production, conversion, distribution, and combustion of these fuels also entails indirect fossil carbon releases related to the inputs of energy at the various processing steps, and inputs of materials in the associated capital goods with embedded energy and carbon release contents, such as cement, steel, and so on. Also, the use of coal, oil, and gas is associated with varying releases of other trace gases (Chapter 3). An accurate comparison of the greenhouse impacts of the different fuels can only be made on a technology-specific, CO_2-equivalent basis. In this chapter, we omit indirect emissions of CO_2. Other trace-gas emissions are taken into account globally (Section B.2).

TABLE 4.2   Carbon Content of Global Fossil Resources

	Oil		Gas		Coal		Total resources	
	< $26/t (1982 $)	total	conv.	conv. &deep	<$40/t (1982 $)	total	conv./ cheaper	all
Resources [1]								
TW-y	132	364	330	2300	560	8535	1022	11199
% of total by fuel	36	100	14	100	7	100	9	100
% of total resources (conv./cheaper)	13		32		55		100	
(all)		3		21		76		100
Carbon content [2]								
BtC	79	217	142	989	426	6495	647	7701
% of total C (conv./cheaper)	12		22		66		100	
(all)		3		13		84		100
1985 consumption								
TW-y	3.97		1.91		2.76		8.6	
Percent of total	46		22		32		100	
1985 carbon release [3]								
BtC	2.18		0.85		2.17		5.2	
Percent of total	42		16		42		100	
Years to exhaustion								
at 1985 rate of consumption	33	92	173	1204	203	3092	118	1296
Change in CO2 level from [4] **complete combustion**								
Absolute (ppm)	20	56	37	258	111	1690	168	2004
relative to 1880-1985 increase (Index 60 ppm = 1.00)	0.34	0.94	0.62	4.29	1.85	28.17	2.81	33.40

[1] Resource data from Haefele et al. (1981). Deep gas includes estimate of U.S. resources only.
[2] From Table 4.1.
[3] Based on Marland et al. (1988), excluding cement.
[4] Assuming 55% atmospheric retention of carbon dioxide and the conversion factors of Table 4.1.

Terawatt-years. One TW-yr is a large energy unit equivalent to about one-tenth of current commercial world energy consumption. The table also indicates the current rate of consumption of these fuels, and how long it would take to completely exhaust them at 1985 rates of consumption.

In total, fossil fuel resources as estimated by Haefele et al. (1981) contain about 8000 billion tons of carbon. Eighty percent of this total is stored in the world's coal resources, which constitute about 76 percent of the energy content of fossil resources.

Only about 9 percent of the world's fossil resources are estimated to be recoverable at the cost ceilings shown in Table 4.2. Here, energy in coal constitutes only about 55 percent of the total, but still about two-thirds of the total carbon content.

Current world fossil fuel consumption is about 46 percent oil, 22 percent gas, and 32 percent coal (see also Chapter 5). If we look at the annual carbon contributions, the figures are 42 percent for oil, 16 percent for gas, and 42 percent for coal. Again, coal contributes a disproportionately large share to carbon releases.

2. Atmospheric Retention of Carbon Dioxide Releases

An important complexity in the discussion of this chapter is the determination of the airborne fraction. In the past, two major approaches have been used (Trabalka 1985). One is the application of a constant airborne fraction (CAF). The other employs carbon-cycle models to calculate the airborne fraction as a function of the emission path, that is, both the amount and timing of releases. In the discussion below, we use both approaches and examine the degree to which they give divergent results for different carbon release trajectories.

Constant Airborne Fraction (CAF). The constant airborne fraction approach is based on interpretations of observed atmospheric measurement data, such as the Mauna Loa records, and estimates of net annual carbon dioxide releases. These data suggest an airborne fraction of anywhere from about 50 percent to about 58 percent for carbon releases from fossil fuels (Perry 1984; Trabalka et al. 1985).

If biospheric releases are accounted for (see Chapter 3), the airborne fraction from observed CO_2 concentrations decreases to about 34–43 percent, depending on the estimate for these releases. In accordance with the discussion in Chapter 3, we assume that afforestation and reforestation activities will over time maintain a neutral biosphere. With this assumption, we can limit ourselves to the range of 50–58 percent for the airborne fraction. We use an intermediate value of 55 percent in our CAF calculations (Tables 4.1 and 4.2). On that basis, 1 billion tons of carbon release results in a 0.26 ppm increase in atmospheric carbon dioxide concentrations. The matrix in Table 4.1 shows conversion factors for going from TW-yr for a specific fuel to parts per million (ppm).[2]

Carbon-Cycle Models. The large-scale release of fossil carbon dioxide represents not only a perturbation of the atmosphere, but an intervention into a complex biospheric system comprising the land biota, oceans, and other carbon-cycle reservoirs, as shown in Figure 3.2. Carbon-cycle (CC) models allow a sophisticated treatment of these perturbations. For example, rates of absorption of CO_2 into the ocean are treated on the basis of chemical and physical measurements. The dynamic feedback effects from climate (and ocean) warming are taken into account. Though significant modeling uncertainties persist, car-

---

[2]The ppm figures are volume-based (ppmv).

bon-cycle models are considered the more reliable tool for predicting the dynamic range of CO_2 levels (Trabalka et al. 1985).

A problem with the CC models is that they can produce significantly different results depending on the model structures and parameters used. Also, empirical data for calibrating the models are only available from periods of rising CO_2 releases, not from periods of consistently declining emissions. The divergence among models becomes particularly pronounced at low carbon release rates (Laurmann and Spreiter 1983)—precisely the conditions that are important during fossil phaseout. Increasing divergence occurs because with declining release rates, the net change in atmospheric concentration becomes a small difference between two comparatively large numbers: carbon emissions are increasingly compensated for by sinks. For the same reasons, it is difficult to predict with certainty what residual levels of CO_2 releases would result in negligible changes to CO_2 concentrations.

We use two alternative carbon-cycle models in our calculations that represent, in terms of outputs for CO_2 concentrations, opposite ends of the spectrum of published models (Siegenthaler 1983).[3] One is a box-diffusion model equivalent to the one developed by Oeschger et al. (1975). As the name implies, the uptake of carbon dioxide in the oceans is modeled as a diffusive process (see Section B.4 in Chapter 2 for details). Based on this model, as much as 75–80 percent of the 1985 fossil carbon releases would have to be eliminated before CO_2 concentrations would no longer rise. The Oeschger box-diffusion model is considered to best reflect the current understanding of geophysical dynamics.[4] It was used in the climate-warming calculations of Chapter 2.

The other model is an outcrop-diffusion model (Siegenthaler 1983) in which nondiffusive processes in the polar regions enhance CO_2 uptake by the oceans. Based on this model, up to almost half of the 1985 level of carbon releases could be tolerated without increases in atmospheric carbon concentrations, and requirements for phasing out fossil fuels would be correspondingly less stringent.

---

[3]The CC model calculations shown in this chapter were kindly provided by Mr. Lashof of the U.S. Environmental Protection Agency (EPA), using a personal computer implementation of the models developed by P. Moore.
[4]For example, Hasselmann (1989) found in a comparative evaluation of conventional CC models that the Oeschger et al. (1975) model best approximated the three-dimensional carbon-cycle model developed at the German Climate Computing Center in Hamburg.

The outcrop-diffusion model is less compatible with current under-standings of the various geophysical and geochemical processes. It does reduce the "missing sink" phenomenon discussed in Section D in Chapter 3, but the unexplained sink in the biosphere could be a transient phenomenon reflecting ongoing adjustments of the biosphere to climatic perturbations that occurred a long time ago. Because of this possibility, the reduction of the "missing sink" in the outcrop-diffu-sion model does not necessarily enhance its credibility. We assign less weight to this model but use it for purposes of sensitivity analysis. The differences between the two diffusion models and the CAF approach will be illustrated in Section D (below).

3. Approximate Correlation between CO_2 Concentrations and Fossil Resource Exhaustion

To get an order-of-magnitude impression of the relationship between fossil resource consumption and atmospheric carbon dioxide levels, it is sufficient to analyze figures based on the simple model of a constant airborne fraction. As shown in Table 4.1, a 1-ppm change in atmo-spheric CO_2 concentrations is roughly equivalent to 3.85 billion tons of carbon release, assuming a 55 percent retention factor.

Table 4.2 shows by how much atmospheric concentrations would increase, based on a 55 percent AF, if all fossil fuel resources were burned. As can be seen from the table, the complete consumption of only the cheaper oil and conventional gas resources would increase atmospheric levels to about 400 ppm, compared to 349 ppm in 1985. The complete use of only the cheaper coal resources would bring atmospheric levels to about 460 ppm. And the complete combustion of all oil, all conventional gas, and all cheaper coal would raise con-centration to the neighborhood of 520 ppm.

To appreciate fully the impact of future fossil fuel consumption on atmospheric composition, ppm increases should be compared to the ppm increase that has occurred since industrialization began. These percentages are shown in the last row of Table 4.2. For instance, with the assumed AF of 55 percent, the combustion of all cheaper oil and coal and all conventional gas reserves would increase atmospheric concentrations to 517 ppm, up by 168 ppm from the 1985 level. The 1985 level, in turn, was about 60 ppm higher than the preindustrial level of about 290 ppm.[5] Relative to this increment of about 60 ppm,

---

[5]There is some uncertainty in the preindustrial concentration level. Within the range of currently preferred values of 280–290 ppm, we use the high-end figure.

complete combustion of these fossil fuels would thus roughly triple the impact on atmospheric composition.

Since we are interested in climatic ceilings on fossil fuel consumption, we examine the fraction of remaining total oil, conventional gas, and total coal resources that could still be burned under each of several carbon budget and ppm levels (abundances). The degree of exhaustion of each fuel category depends on the mix of fuels burned. Table 4.3 gives the percentage of existing resources that would be consumed for a given fuel mix and carbon release budget. Also shown are the approximate ppm levels and concentration increases associated with these carbon budgets, again based on a simple CAF calculation.

The upper end (about 800 btC or about 550 ppm) of the range of abundances shown in the table corresponds to a near doubling of the preindustrial atmospheric CO_2 level. It indicates the point for which severe disruption of the climate has been traditionally predicted. The lower end (200–300 btC or about 400 ppm) denotes an approximate threshold level for severe climate disruption when the findings of Chapter 2 are taken into account (see also Section D below). Finally, Table 4.3 shows how much energy would be obtained under the indicated ppm levels and carbon budgets, assuming different fossil fuel mixes.

D. GLOBAL FOSSIL CARBON BUDGET

1. Carbon-Cycle Modeling of Fossil Fuel Burn Trajectories

In Chapter 2, we used the warming impacts of three energy scenarios that collectively cover the span of published projections to set overall benchmarks. In the following analysis, we use the same span of energy scenarios to illustrate the quantitative differences among the three carbon-cycle models as a function of the emission scenario. At the same time, we abstract from the particulars of the scenarios used in Chapter 2 and take a closer look at alternative fossil carbon trajectories.

Below, we define six scenarios, or trajectories, of global fossil CO_2 releases for the period between 1985 and the year 2145. Each of the six trajectories represents a certain "trend paradigm" that is likely to be relevant for certain subperiods. Scenarios A–C test the carbon-cycle models for constant-to-growing emissions. Scenarios D–F represent alternative emission paths for a fossil fuel phaseout.

The key trajectory assumptions for our sensitivity analysis are

TABLE 4.3 Exhaustion of Fossil Resources under Various Carbon Budgets(1)

	Index of energy obtained	Carbon release (BtC)				
		200	300	400	600	800
Atm. CO2 level (ppm) (2)		401	427	453	505	557
Increase over 1985 level (ppm)		52	78	104	156	208
Fossil fuel mix						
As in 1985						
Energy obtained (TW-y)	1.00	341	511	682	1022	1363
Resource exhaustion						
Cheaper oil		117%	175%	234%	350%	467%
total oil		42%	64%	85%	127%	169%
conventional gas		31%	47%	62%	94%	125%
total gas		4%	7%	9%	13%	18%
cheaper coal		15%	22%	30%	45%	60%
total coal		1%	1%	2%	3%	4%
All gas						
Energy obtained (TW-y)	1.37	465	698	931	1396	1862
Resource exhaustion						
conventional gas		141%	212%	282%	423%	564%
total gas		20%	30%	40%	61%	81%
All oil						
Energy obtained (TW-y)	0.98	335	503	671	1006	1342
Resource exhaustion						
cheaper oil		254%	381%	508%	762%	1017%
all oil		92%	138%	184%	276%	369%
All coal						
Energy obtained (TW-y)	0.77	263	394	525	788	1051
Resource exhaustion						
cheaper coal		47%	70%	94%	141%	188%
total coal		3%	5%	6%	9%	12%

(1)Shown as the percent of existing resources (Table 4.2) that would be consumed for a given C release budget and resource mix.
(2)Based on 55% atmospheric retention and the resources shown in Table 4.2.

A. *An average growth rate in CO_2 releases of 2 percent (annually compounded).* This rate is roughly equivalent to the carbon releases from fossil fuel growth as projected in the two major conventional world energy studies (WEC [1978] and Haefele [1981]). These studies underemphasize efficiency improvements and project Third World development to follow the energy-intensive growth patterns seen in industrialized countries during earlier phases of their development. From today's viewpoint, path A should be seen as an upper limit for the range of plausible fossil CO_2 release rates.

B. *Constant 1985 per capita emissions.* This trajectory illustrates how population dynamics alone would drive global emissions if all other factors remained constant. It is assumed that world population would stabilize at about 7.8 billion people by the middle of the next century. This is equivalent to a 60 percent increase in world population over 1985 levels and corresponds to United Nations' "Low" forecast (U.N. 1982; 1984; see Table 5.2).

C. *Constant global emissions at the 1985 level.* This trajectory extrapolates the trend of global emissions in the early 1980s, which was reasonably flat (see Fig. 5.4). This trajectory corresponds to the lower end of the range of fossil burns typically discussed in official policy deliberations.

D. *Average linear decline of global emissions at a rate of -2.0 percent for 40 years, then constant global emissions at 20 percent of the 1985 level.* This trajectory explores the impact of reducing emissions by a major fraction early on, while fossil fuels would continue to be used at a significant residual level in late decades.

E. *Average linear decline of 1985 global emissions to zero at a rate of -1.0 percent per year.* This trajectory explores the implications of a more gradual fossil fuel reduction based on a 100-year target for a complete phasing out. It shows the impact of a slower speed of CO_2 reductions over the next few decades relative to the more aggressive approach of D, followed, however, by a complete fossil phaseout in later decades.

F. *Delayed phaseout composite.* In this scenario, it is assumed that the onset of fossil phaseout is delayed due to the difficulties of turning around energy policies and trends on an international scale. There is about a decade of continued fossil growth (as in A), followed by a decade of approximately constant fossil fuel use (as in C), and return to 1985 levels within about 25 years. In the following 30 years, releases decline to about 20 percent of the 1985 level, and then remain constant at that level (as in D).

Table 4.4 shows the total cumulative carbon releases along each trajectory for the period 1985–2100. Scenarios A–C result in cumulative releases of about 2400 btC, 900 btC, and 600 btC, respectively.[6] The three phaseout trajectories also differ significantly. Whereas trajectory D releases only about 200 btC, trajectory E results in about a 260 btC release, and scenario F in a 300 btC release for 1985–2100.

---

[6]In accordance with our treatment of biotic and other carbon releases in Chapters 2 and 3, the above scenarios are defined to reflect carbon releases from fossil fuels.

TABLE 4.4   Carbon Dioxide Level as a Function of Carbon Release Trajectories[a]

CO2 concentration (ppm)				400	450	500	550
Index 1985 ppm (349=1.00) Index preindustrial (290ppm=1.00) *Change rel. to 1860-1985 ppm increase (60 ppm=100%)*				1.15 1.38 *85%*	1.29 1.55 *168%*	1.43 1.72 *252%*	1.58 1.90 *335%*
Scenario	1985-2100 Cumulative C releases btC	C release rate (1985=1.00) 2025	C release rate (1985=1.00) 2085	**Year in which level is reached**			
A. +2% compounded abs. CAF Box-Diffusion Outcrop-Diffusion	2402	2.21	7.24	 2013 2010 2015	 2030 2026 2032	 2043 2037 2045	 2053 2046 2055
B. Constant per cap CAF Box-Diffusion Outcrop-Diffusion	904	1.25	1.28	 2015 2012 2020	 2040 2034 2050	 2062 2055 2085	 2085 2076 2126
C. Constant absolute CAF Box-Diffusion Outcrop-Diffusion	610	1	1	 2022 2020 2037	 2059 2057 2103	 2095 2095 2177	 2132 2133 –
D. -2% linearx40, const. CAF Box-Diffusion Outcrop-Diffusion	205	0.2	0.2	 2094 – –	 – – –	 – – –	 – – –
E. -1%linear CAF Box-Diffusion Outcrop-Diffusion	262	0.6	0	 2036 – –	 – – –	 – – –	 – – –
F. Delayed Phase-out CAF Box-Diffusion Outcrop-Diffusion	301	0.5	0.2	 2021 2016 –	 – – –	 – – –	 – – –

[a]Box-diffusion model and outcrop-diffusion model from Siegenthaler (1983).

We investigate three basic questions:

1. By when, given the above trajectories of fossil CO_2 release, would atmospheric CO_2 concentrations exceed the 400-ppm limit?
2. What is the upper limit for cumulative fossil carbon releases under the 400-ppm limit?
3. How divergent are the answers obtained from CAF and CC model calculations?

The results of our calculations are presented in Figure 4.1a–f and in Table 4.4.

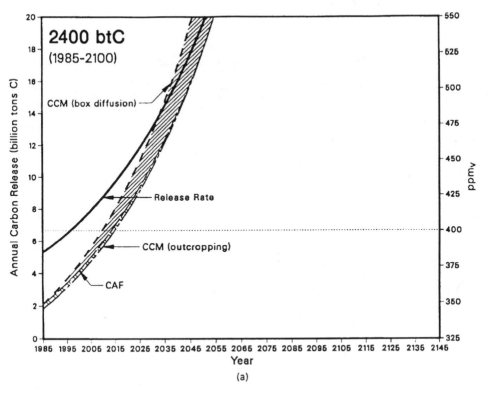

Figure 4.1a CO_2 *concentration — trajectory A.*

2. Viable Global Carbon Trajectories under a 400-ppm Limit

Inspection of Figures 4.1a – f and Table 4.4 reveals that under a 400-ppm limit for atmospheric carbon dioxide concentrations, only scenarios D – F are viable.[7] In scenario D, CO_2 concentrations remain safely below the limit, despite a residual carbon release of about 20 percent of the 1985 level. In scenario E, CO_2 levels peak around 2040 to 2060 and then decline. Only for the box-diffusion calculation do they exceed the 400-ppm limit, and then only temporarily and by a few ppm. Along trajectory F, a similar peak is observed earlier, around 2020 – 2030. Here, concentration levels stabilize over the longer run. Again, the box-diffusion model predicts levels slightly above 400 ppm.

For trajectories A and B, the 400-ppm limit would be reached somewhere between 2010 and 2020. In scenario C, this threshold would be exceeded by the middle of the century or later.

[7]We base the following discussion on the results obtained from the CC models and ignore the CAF results (see also Section B).

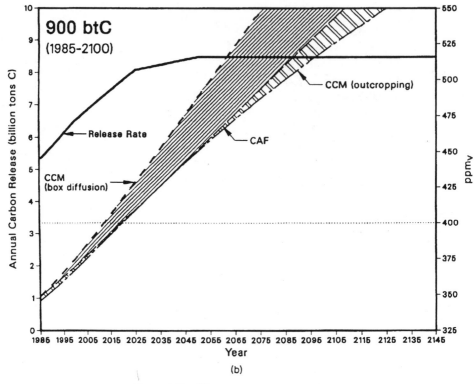

Figure 4.1b *CO$_2$ concentration — trajectory B.*

Early versus Delayed Phaseout. The key interpretations of these results are as follows:

- The time gained by moving from high growth to a constant level of CO$_2$ releases is much smaller (on the order of a few years to a few decades) than the time gained by moving from a constant level of CO$_2$ release to negative growth (of the order of decades to centuries).
- Among trajectories of declining CO$_2$ emissions, a strategy that aggressively seeks reductions early on (scenario D) provides much more time to develop and deploy new fossil-substituting technologies than one in which reductions are achieved at a more lackadaisical pace (scenarios E and F).

Put another way (Lovins et al. 1981),

- A unit of CO$_2$ reduction achieved in the near term is worth several units of CO$_2$ reductions achieved in the longer term.

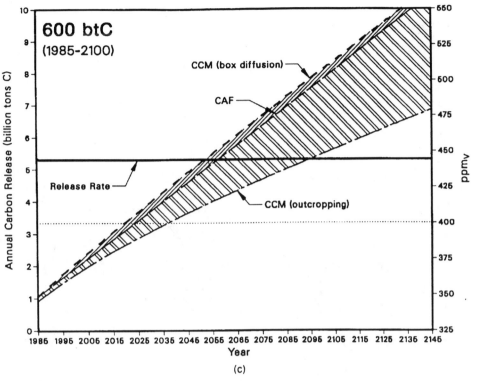

Figure 4.1c *CO$_2$ concentration — trajectory C.*

These findings again underline that the opportunity cost and risk of not reducing release rates early on grow disproportionately with each year of delay.

CC Models versus CAF. When looking at the difference between carbon-cycle models and CAF calculations, the following observations can be made:

- For trajectories of exponential growth (A), the difference between CAF and CC models is minimal.
- For the phaseout scenarios (D–F), the CAF approach overestimates carbon dioxide concentrations. The divergence grows as low release rates are reached. At such low levels of release, carbon sinks begin to overwhelm sources, resulting in an effective stagnation or slow decline of CO$_2$ concentrations (Figs. 4.1*d–f*). This effect is not captured by the CAF model.

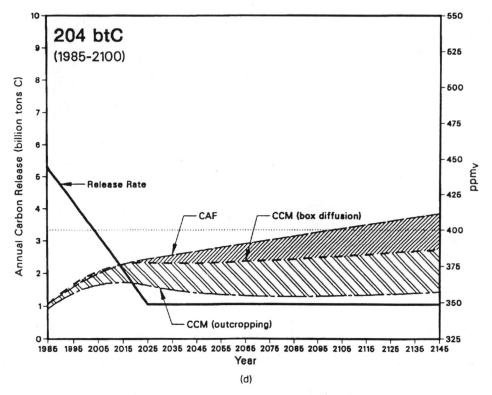

Figure 4.1d *CO_2 concentration — trajectory D.*

- For the phaseout scenarios, the difference in CO_2 concentrations between the box-diffusion and the outcrop-diffusion model is significant. By 2100, it is 27–36 ppm.

To illustrate the latter point, consider trajectory F. The peak concentration in the box-diffusion calculation is 407 ppm, which is reached in about 2030. In the outcrop-diffusion calculation, the peak is 387 ppm, or 20 ppm lower. In 2100, the box-diffusion model calculates a CO_2 concentration of 404 ppm. According to the outcrop-diffusion model, CO_2 levels would have declined to 368 ppm by then, for a difference of 36 ppm.

3. Global Fossil Carbon Budgets under a 400-ppm Limit

Inspection of box-diffusion trajectories D–F reveals that a 400-ppm limit would require a restriction of carbon releases to about 300 billion tons between 1985 and the year 2100. Note also that the exact figure

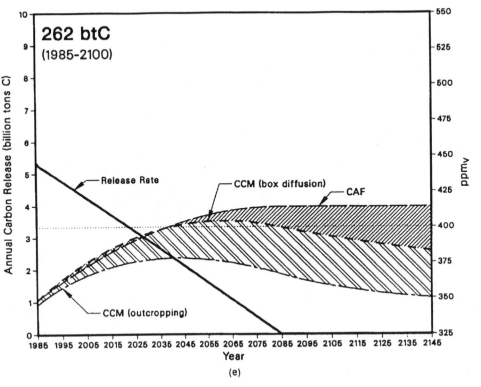

Figure 4.1e *CO₂ concentration — trajectory E.*

depends on the path of the fossil phaseout. For the same concentration limit, the phaseout scenario F allows a 50 percent greater amount of carbon release than trajectory D, and provides a 15 percent greater carbon budget than trajectory E (see Table 4.4).

Because the CAF approach assumes increases in atmospheric concentrations at all levels of carbon release, this model underestimates the allowable carbon budget under a 400-ppm limit by up to about 100 btC relative to the box-diffusion model (see Table 4.3). The outcrop-diffusion model, on the other hand, extends the upper limit for the carbon budget to about 450 btC.[8]

Because the outcrop-diffusion model is considered less credible, and

[8]See Svennigsson (1984), who calculated carbon dioxide concentrations for the World Energy Conference, IIASA, and Goldemberg et al. global energy scenarios using the outcrop-diffusion model. A 450-btC budget (1985–2100) is obtained, for example, if one follows the Goldemberg et al. scenario until 2020 and then linearly phases out all but about 1 btC of fossil fuels by 2100.

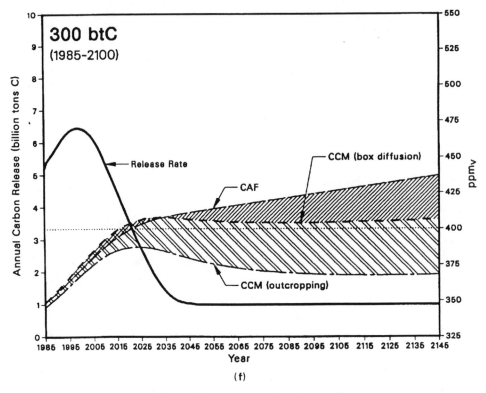

Figure 4.1f *CO₂ concentration — trajectory F.*

in light of recent measurements indicating that airborne fractions may be rising, we conclude that

- The upper limit for cumulative fossil carbon releases between 1985 and 2100 should be about 300 btC.

At this stage in our analysis, this budget is a working hypothesis. In our subsequent analyses in Chapters 5 and 6, we investigate its practical feasibility, and if need be, make iterative adjustments. In this exploration of practical phaseout schedules, the 450-btC figure is used for sensitivity analyses.

4. Fossil Energy Budget under a 300-btC Limit

A carbon budget of 300 billion tons is equivalent to 511 TW-yr of energy at the 1985 fuel mix (Fig. 4.2). Using the index of energy obtained for various other fuel mixes (Table 4.3), we can calculate that

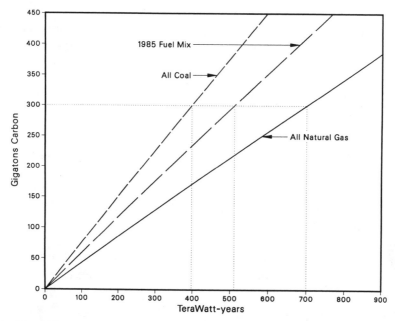

Figure 4.2 *Global carbon and fossil energy budgets under a 400-ppm atmospheric CO_2 limit.*

under this budget the world would be able to obtain some 700 TW-yr of energy if all consumption were gas, or some 400 TW-yr if all consumption were coal.

Based on these figures, we can translate the global carbon budget into the number of years it would take to consume this budget at 1985 levels of fossil fuel consumption (Fig. 4.2). With a constant fuel mix, the 300 billion t budget would last about 57 years, or until the year 2042. Depending on the direction of change in global fossil fuel mix, this period might realistically shorten or lengthen by a decade or so.

5. Allowable Depletion of Fossil Fuel Resources

A 300-btC budget means major restrictions on the use of global fossil resources. The quantitative implications for alternative fuel mixes can be seen from the data in Table 4.3. At the 1985 mix of fuels, the consumption of fossil resources would be limited as follows:

- No more than 64 percent of total oil resources.
- No more than 47 percent of conventional gas resources.
- No more than 22 percent of the cheaper coal resources.

These figures clash with the conventional assumption that all conventional oil and gas resources would probably be consumed before a major shift away from fossil fuels would occur. Our analysis suggests that

• Climate stabilization requires keeping significant portions of even the world's conventional fossil resources in the ground.

Such a requirement is a stark contradiction to all conventional energy planning and illustrates the magnitude of the greenhouse challenge.

Of course, some interfossil fuel shifting is feasible and desirable to get the most energy out of the global carbon budget. But the room for such cross-substitution is limited by the structure of fossil resources. While it would be desirable from a resource efficiency point of view to shift fossil fuel consumption to more plentiful (and often cheaper) coal, both the greenhouse effect and other environmental problems of coal use point in the opposite direction, toward gas and oil. Since gas resources are the second largest after coal, the need to shift away from oil and from carbon-intensive coal is perhaps best accommodated by cross-substitution with gas. Fundamentally, however, a much more radical shift away from all fossil fuels is needed than has been anticipated so far.

6. Viable Energy Scenarios under a 400-ppm Limit

The calculations in Chapter 2 had already indicated that climate stabilization is incompatible with conventional world energy scenarios, which are based on rising fossil fuel consumptions. This finding is further clarified by the calculations of this chapter. In particular, the most widely quoted world energy scenarios, that is, the "IIASA Low" and World Energy Conference "low" trajectories, which are approximated by Figures 4.1a and b, would "blow" the 400-ppm CO_2 ceiling by 2010–2025. This result is obtained no matter what carbon-cycle model is used, and no matter what assumptions are made about fossil fuel use in the decades beyond 2020–2030 which are not covered by these scenarios. Simply put, conventional world energy strategies and projections need to be scrapped if climate stabilization is to be achieved.

Our more detailed exploration in this chapter further reveals that even constant carbon releases at 1985 levels pose unacceptably high climatic risks. This type of scenario had been proposed by Goldemberg et al. (1988), who argued that investments in efficiency improvements could keep energy use flat but that the world would likely use up all its oil and gas reserves before a switch to renewables would occur. Given the magnitude of these resources (see Table 4.2), fossil fuels would under this assumption completely dominate energy use for the remainder of the twenty-first century.

This outcome is again incompatible with climate stabilization, irrespective of the carbon-cycle model assumed. The Goldemberg et al. scenario would obey the 400-ppm limit if it is assumed that beginning in about 2020–2030, a linear fossil fuel decline to a release rate of 1 btC/yr is achieved by 2100. Even then, compatibility is achieved only with the outcrop-diffusion model and not with the more plausible box-diffusion model. Greater climatic acceptability would depend on levels of reforestation and other trace-gas control that are greater than those assumed in Chapter 3.

Finally, we see that the scenario by Lovins et al. (1981), with a cumulative carbon release of only about 150 btC between 1985 and 2100, would represent a substantial safety margin against climatic surprises for the worse. This margin could buffer against stronger-than-expected feedback effects, or against serious slips in reforestation efforts and other measures related to nonfossil carbon trace gases. This scenario would be compatible with the strict 300-btC budget of the box-diffusion model even if, following Goldemberg et al., no fossil fuel substitution were to occur. Then again, the full realization of this scenario, which requires implementing by 2030 most efficiency potentials and displacing 80 percent of fossil fuels with renewables, has already been preempted due to the inertia of recent and current energy policies.

These findings can be summarized as follows:

- The conventional world energy scenarios (World Energy Conference and IIASA) are incompatible with climate stabilization.
- The range of desirable world fossil fuel trajectories is circumscribed by the scenario of Lovins et al. at one end, and a renewables-based extrapolation of the scenario of Goldemberg et al. at the other end. The former can be seen as representing the outer limit of logistic feasibility, while the latter marks the outer limit of climatic acceptability.

REFERENCES

Goldemberg, J., T. B. Johannson, A. K. N. Reddy, and R. H. Williams (1988), *Energy for a Sustainable World*, Wiley, New York.

Haefele, W., et al. (1981), *Energy in a Finite World*, Ballinger, Cambridge, MA.

Hasselmann, K. (1989), Personal communication, Max-Planck Institut für Meteorologie, Hamburg.

Laurmann, J. A., and J. R. Spreiter (1983), "The Effects of Carbon Cycle Model Error in Calculating Future Atmospheric Carbon Dioxide Levels," *Climatic Change*, Vol. 5, No. 3, pp. 145–181.

Lovins, A. B., L. H. Lovins, F. Krause, and W. Bach (1981), *Least-Cost Energy: Solving the CO_2-Problem*, Brick House, Andover, MA. (German version: *Wirtschaftlichter Energieeinsatz: Losung des CO_2-Problems*, Karlsruhe, FRG, 1983.)

Marland, G., and R. M. Rotty, (1984), "Carbon Dioxide Emissions from Fossil Fuels: A Procedure for Estimation and Results 1950–82," *Tellus*, Vol. 36B, No. 4 (Sept.), pp. 232–261.

Marland, G., et al. (1988), Estimates of CO_2 Emissions from Fossil Fuel Burning and Cement Manufacturing Using the United Nations Energy Statistics and the U.S. Bureau of Mines Cement Manufacturing Data, Carbon Dioxide Information Analysis Center (CDIAC), Numeric Data Collection, Oak Ridge National Laboratory, Oak Ridge, TN.

Oeschger, H., et al. (1975), "A Box Diffusion Model to Study the Carbon Dioxide Exchange in Nature," *Tellus*, Vol. 27, pp. 168–192.

Perry, A. M. (1984), *Atmospheric Retention of Anthropogenic CO_2: Scenario Dependence of the Airborne Fraction*, EPRI/EA/3466, Electric Power Research Institute (EPRI), Palo Alto, CA.

Siegenthaler, U. (1983), "Uptake of Excess CO_2 by an Outcrop-Diffusion Model of the Ocean," *Journal of Geophysical Research*, Vol. 88, pp. 3599–3608.

Svennigsson, P. (1984), *Global Fossil Energy Use in a CO_2 Context*, Status Report 1984–08-21, Environmental Studies Program, University of Lund, Sweden.

Trabalka, J. R., ed. (1985), *Atmospheric Carbon Dioxide and the Global Carbon Cycle*, DOE/ER/0239, U.S. Department of Energy, Washington, D.C., Dec.

Trabalka, J. R., et al. (1985), "Human Alterations of the Global Carbon Cycle and the Projected Future," in J. R. Trabalka, ed., *Atmospheric Carbon Dioxide and the Global Carbon Cycle*, DOE/ER/0239, U.S. Department of Energy, Washington, D.C., Dec.

U.N. (1982), *Population Bulletin of the United Nations*, Department of International Economic and Social Affairs, United Nations Secretariat, New York, No. 14.

U.N. (1984), *Population Bulletin of the United Nations*, Department of International Economic and Social Affairs, United Nations Secretariat, New York.

WEC (1978), *World Energy Resources, 1985–2020*, Guildford, U.K., IPC Science and Technology Press for the World Energy Conference.

Part 2

Toward a Global Compact on Climate Stabilization and Sustainable Development

In Part Two, we discuss policy issues related to the formulation of an international agreement to stabilize the world's climate.

Chapter 5 develops quantitative formulations for dealing with the equity issue between industrialized and developing countries, either through the physical allocation of the fossil carbon budget or through compensatory financial arrangements.

Chapter 6 develops separate fossil carbon reduction milestones for developing and industrial countries as a group. We explore what rates of converting capital stocks would be needed to meet climatic imperatives, and evaluate these rates in terms of both logistic constraints and factors of political economy.

Chapter 7 translates the analysis of Chapters 1–6 into suggestions for shaping a global climate convention. Emphasis is placed on the use of a target-based approach using a global fossil carbon budget and on mechanisms that would transfer capital and technologies to the developing nations.

5

How Could the Global Carbon Budget Be Shared?

A. CLIMATE CHANGE AND GLOBAL EQUITY

One important insight driven home by the greenhouse effect is that gross international inequity and ecological maintenance are fundamentally incompatible. In fact, the global environmental crisis and the warming threat could bring about a realignment of international policies that goes far beyond the control of greenhouse gases in the narrow sense (Reddy 1989): historically, industrialized countries have for the most part allied themselves with Third World elites at the expense of basic-needs-oriented development. In view of the global environmental consequences, these alliances have now become anachronistic. Solutions to the global warming problem and related environmental threats will necessitate satisfying basic needs everywhere.

Climate stability is a limited resource: the global climate can tolerate only a certain amount of perturbation before it will irreversibly deteriorate. This climatic resource takes the form of a buffering capacity against greenhouse gas releases. In this text, we have quantified this resource in the form of a global budget for remaining allowable fossil carbon releases. We now deal with ways of allocating the global carbon budget in an equitable manner. Past discussions on the issue of sharing the limited fossil fuel resources of the earth have usually been based on the assumption that these fuels offer a unique developmental benefit, and that this benefit grows in proportion to the amount of fossil fuels used. This common assumption has led to sometimes acrimonious debates over an equitable physical distribution of cumulative fossil fuel consumption among the world's nations.

Our analysis in this book suggests a very different view: that from here on in, there is a relatively modest optimum amount of cumulative fossil fuel use beyond which these fuels become not only environmentally disruptive but economically costly as well. Most industrialized countries find themselves in this economic and environmental trap already, and will be foregoing significant welfare increments until they greatly reduce fossil fuel use and related pollutant emissions.

Developing nations have a unique opportunity to leapfrog into economies based on largely nonfossil energy systems. Such systems would rely on combinations of capital-saving high-efficiency technologies and low-pollution supply technologies based on renewables. They potentially could provide larger economic and developmental benefits than intensive fossil fuel use patterned after the evolution of industrialized countries.

From this perspective, the equity issue loses much of its divisiveness. Yet under a tight climate-stabilizing global fossil carbon budget, the equity issue does retain its importance insofar as most or all of this limited budget will be needed to support the global transition to largely nonfossil energy systems. Fossil fuels thus offer a vital *transitional* benefit (see Section B and also Chapter 6).

The basic investigation in this chapter is whether and how these transitional benefits of fossil fuel use could be obtained by both industrialized and developing countries, using an equitable physical allocation of the global fossil carbon budget as far as feasible, combined with compensatory capital and technology transfer programs. We therefore investigate two topics:

1. What might be an equitable yet feasible physical allocation of the global carbon budget?

    How much more fossil fuel consumption would industrial countries (ICs) be allowed if climatic resources were shared equitably?

    Conversely, we might ask: How much more fossil fuel would developing countries (DCs) be entitled to burn if the global fossil carbon budget were to be apportioned according to strict equity criteria?

    Could ICs manage their transition to nonfossil energy systems if limited to a strictly equity-based portion of the global carbon budget?

How would Third World countries be affected if they were to receive less than their calculated fair share?
2. If strict allocational equity is not feasible on a physical level, how could sustainable Third World energy development be ensured anyway?

How could DCs be compensated if they were to forego the full use of their equity-based fossil carbon budget?

How should the burden of compensatory assistance to the Third World be shared among industrialized countries?

Section B develops quantitative equity criteria for the physical allocation of the global carbon budget. Sections C and D apply these equity formulas to historic and future carbon dioxide releases. Section E deals with compensatory arrangements that could flexibly complement a less-than-perfect physical allocation of the fossil carbon budget.

B. HOW COULD EQUITY GOALS BE QUANTIFIED?

A logical principle for assigning responsibility for climate stabilization would be that those who cause the damage pay for the cost of abatement. This principle is by now the basis of environmental legislation in many nations. If fully enforced, it ensures that the burden of environmental repair will be carried equitably. In internalizing previously ignored costs, it also sends the right economic signals to the producers of pollution and to the consumers of products associated with such pollution.

Identifying the principal culprits in the destruction of the world's climate is not difficult. The history of greenhouse gas emissions shows that the industrialized countries have been and continue to be by far the major source of climate destabilization: they overwhelmingly dominate CFC emissions, clearly dominate carbon dioxide emissions, and contribute close to half of global anthropogenic methane releases (see Chapter 3 and below). Put another way, the prevailing lopsided international *economic* order is reflected in a less-recognized but similarly inequitable *ecological* order that is starkly revealed by the global climate threat. The main focus in this chapter is therefore on an equitable distribution of the global fossil carbon budget among developing and industrialized regions as a whole. Of course, important allocational issues also exist among nations and groupings within each of those two categories. These are further investigated in Chapter 6.

1. Global Interdependence: Accounting Issues in Measuring Equity

Developing quantitative distributional criteria for allocating climate-stabilization responsibilities is inherently difficult. Besides the challenge of including all the greenhouse gases in such allocations, rigid equity criteria run the risk of creating artificial accounting boundaries in a world that is highly interdependent.

For example, one might argue that developing countries have in fact contributed to climate change in a major way through deforestation driven by population growth, if not yet through fossil fuel use or CFC releases. This is, of course, true. In fact, the fossil carbon budgets developed in Chapter 4 are explicitly based on maintaining biotic carbon storage at mid-1980s levels. Given current deforestation trends, this will involve a major afforestation and reforestation campaign. The bulk of this campaign will need to be carried out by developing countries.

On the other hand, ecological destruction in the Third World is inextricably linked to the history of colonization, as well as to persistent postcolonial patterns of inequitable trade. Over the last 200 years, industrialized countries have been actively and often forcefully integrating Third World regions into their economic cycles. In the course of this integration, ICs also exported their own model of economic and social development.

Today, many people in developing countries aspire to obtain at least some of the consumer goods prevalent in the wealthy countries. As a result, DCs are also displaying a growing demand for fossil fuels and other greenhouse-active products. But now that the risk of climate destruction has become broadly recognized, it turns out that much of the biosphere's buffering capacity against trace-gas emissions has already been consumed.

Arguably, industrialized countries have primary responsibility for climate stabilization not only because of their disproportionate past and present role in altering the global atmosphere, but also because of their disproportionate capabilities to mitigate this threat. Whereas the developing countries are laboring under immense problems of debt, debilitating population growth, and capital scarcity, industrialized nations have the technology, the capital, and the organizational faculties that are needed to "clean up their act." Moreover, their influence on the Third World's investment patterns and resource consumption through lending, technology export, and trade is immense, and could be used to help those countries from growing into a major climate-destroying force of their own.

For these and other reasons, any international agreement that may ultimately be reached might well deviate from simple equity-based formulas. Nor will the economic fate of each country hinge on the exact share it receives of the global fossil carbon budget, or for that matter, on any other greenhouse gas reduction target. The economic history of the world has shown that countries with the most unfavorable natural resource endowments can overcome such handicaps with technological and cultural inventiveness.

The more important prerequisites of economic success seem to be participatory political structures, equitable distributions of land and income, self-reliance, and the freedom to shape development policies according to national priorities rather than those of the economically dominant powers (Senghaas 1977; 1982).

Nevertheless, it is useful to know how the task of climate stabilization would be allocated under strict adherence to international equity criteria. In particular, such an analysis provides countries and regions with a framework for the goals they should set themselves in energy planning so long as formal international agreements have not been reached.

Can Fossil Carbon Releases Be Used as a Proxy for Greenhouse Equity? Figures 5.1–5.3 show the contributions of industrialized and developing countries to the greenhouse problem, using several indices. The first figure shows an estimate from a climate-stabilization analysis by the U.S. Environmental Protection Agency (Lashof and Tirpak 1989) in which emissions of all greenhouse gases were disaggregated

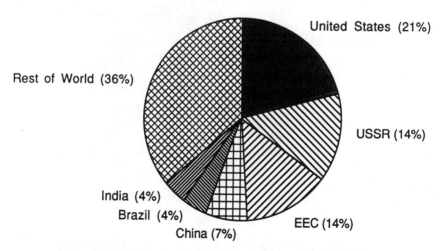

Figure 5.1 *Regional contributions to global warming, 1980s (%).*

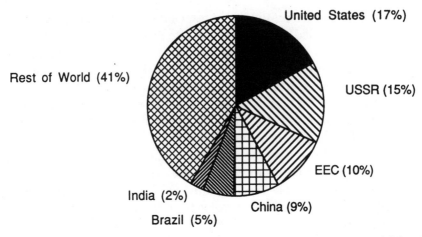

Figure 5.2 *Regional contributions to global carbon releases from fossil fuel use and deforestation, 1980s (%).*

into major countries and regions and then weighted by their respective *warming impacts*. This sophisticated, warming-based analysis can be compared with Figures 5.2 and 5.3, in which percentages are compiled for the same disaggregations, but derived from greenhouse gas *release rates*. Figure 5.2 shows the results for total carbon dioxide releases (fossil releases and deforestation), based on the combined data of Houghton et al. (1987) and Marland et al. (1988). Figure 5.3 offers fossil releases alone.

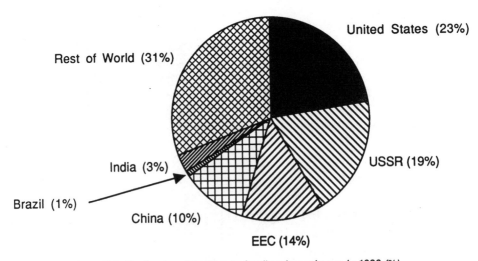

Figure 5.3 *Regional contributions to fossil carbon releases in 1986 (%).*

A comparison of the three figures reveals that the regional distribution of fossil carbon releases could be used as a reasonable proxy for the regional contributions to global warming: the major industrialized countries explicitly shown in the pie charts (the United States, the Soviet Union, and the nations of European Economic Community) contributed 49 percent of the annual global warming increment during the 1980s. Their contributions to fossil carbon releases was about 56 percent. Their share in carbon dioxide releases from fossil fuels and deforestation was 52 percent.

For some countries with large land area and deforestation rates, the difference between these indices can, of course, be dramatic. For example, Brazil accounts for 5 percent of fossil and biotic carbon releases, 1 percent of fossil carbon releases, and 4 percent of the annual global warming increment. The latter figure is mainly the result of the climate-modeling convention to treat the biosphere as neutral (see Chapters 2 and 3).

Note that all of the comparisons in Figures 5.1–5.3 are based on current annual contributions, not cumulative contributions over historic periods. The effect of including historic emissions in equity considerations is illustrated for fossil carbon emissions in Section C.

2. Equity Formulas

In establishing equity formulas, at least three basic alternatives are available: carbon-release-based models, C/gdp-based (fossil carbon intensity of gross domestic product) models, and population-based models.

Carbon-Release-Based Models. In its simplest form, this approach consists of allocating the cumulative fossil carbon budget in proportion to each country's current carbon releases. Each country would have to disengage from the use of fossil fuels at the same average speed (though individual reduction milestones could, of course, vary). Based on 1986 figures, developing countries would obtain a 28 percent share[1] of the global fossil carbon budget, or about 30 years' worth of releases at 1986 rates.

This approach essentially replicates the current status quo. It could be the basis for an acceptable agreement only

- If new nonfossil technologies can provide energy services more cheaply than fossil-based ones;

[1]Based on Marland et al. (1988). See Table 5.4.

- If the potential of these technologies for reducing carbon emissions is sufficiently large in all economies irrespective of their absolute size and stage of development; and
- If all countries have comparable access to the technical know-how, financing, and organizational capacities to implement these technologies.

The first proposition is partially true. It holds for most demand-side technologies. On the supply side, the picture is mixed. At least when externalized costs are ignored, fossil fuels at current prices are significantly cheaper than several of the presently available renewables-based resources. However, this cost differential could be much smaller or of inverse sign in many local settings, particularly in the Third World. And some renewables are already cost-effective against present fossil-based energy supplies.

The second proposition is quite plausible. There is strong evidence that efficiency-based percentage reductions in fossil carbon releases in developing countries could be at least as large as in the industrialized countries, and probably significantly larger: Third World modern-sector capital stocks are consistently less energy-efficient than those in the industrialized countries, and the discrepancy is even more pronounced in the traditional or informal economic sectors (Goldemberg et al. 1988). And on the supply side, the technical potential of renewables is at least as large in the developing countries as in the industrialized countries.

However, the third proposition clearly does not apply at all. Indebtedness, capital scarcity, and lack of domestic technological and administrative capacities are the hallmark of the current Third World predicament. These will hinder not only the implementation of efficiency improvements, but also the timely development and commercialization of advanced renewables-based technologies (see also Chapter 6).

Furthermore, the current narrow selection of fully commercial, cost-effective renewables-based supply technologies is a particular handicap for developing countries. In the near-to-medium term, the high growth rates of energy service needs in these countries cannot be fully offset by converting existing capital stocks to more efficient technologies, as is the case in industrialized countries. Faced with the need to expand supplies, they would obtain major portions of this supply expansion at lowest cost from fossil fuels, especially in nations with accessible domestic resources.[2] On the other hand, a larger-than-expected contribution could already be provided from cost-effective renewables, particularly when foreign exchange considerations and other externalities are taken into account.

On balance, however, conventional least-cost criteria would still lead to increased DC fossil fuel consumption in the near-to-medium term. If percentage reduction milestones were uniformly applied to developing and industrialized countries, developing countries would be forced to make relatively earlier and greater use of higher-cost renewables. On a per-unit basis, the early reduction of fossil carbon releases could thus cost them more than it would cost the industrialized countries.

Industrialized countries, on the other hand, experience relatively modest growth in energy services. They could overwhelm that growth merely by inserting cost-effective efficiency technologies into existing capital stocks through retrofit or replacement investments (see Chapter 6). They could therefore achieve major fossil carbon reductions without increasing the cost of energy services. This would also gain them the necessary time to develop cheaper nonfossil supply technologies fully while lowering fossil fuel consumption in the meantime.

Reduction Milestones Based on C/gdp Ratios. This approach is similar to the release-based model. All countries would be required to reduce their C/gdp ratios by a certain percentage by a given year. This is an improvement insofar as the proposal recognizes that current C/gdp ratios in developing countries are higher than in industrialized countries. The model also would give countries with greater gdp growth rates a greater fossil carbon allowance. This would buffer developing countries against the disproportionate costs of near-to-medium term expansion of nonfossil supplies. However, this model still does not address differences between the ICs and the DCs in the capacity to develop, commercialize, finance, and implement reduction technologies, and the broader international equity questions associated with these differences.

There are other drawbacks. First, it is inherently difficult to relate C/gdp reduction targets to the permissible fossil carbon budget. To do

[2]Some developing countries have significant fossil resources but face very large infrastructure investments if they want to exploit them. The coal resources in mainland China, which would require enormous transportation investments, are but one example of this.

so would require the ability to reliably measure gdp and to make long-term forecasts of regional and global gdp growth. The volatility of gdp projections could require frequent adjustments to the C/gdp reduction targets. Measurement problems and the need for frequent renegotiations could undermine the consensus necessary for ongoing cooperation.

Second, the driving force of fossil fuel consumption is not so much the growth in gdp per se, but changes in the kinds and amounts of energy services employed to produce a unit of gdp. This *energy service intensity of gdp* (Krause 1980; Lovins et al. 1981) is both a function of the country's resource endowment and industrial mix, and of the stage of development. Many developing countries are experiencing phases of economic development in which the energy service intensity of gdp is rapidly growing — a trend that was also experienced during the earlier development stages of currently industrialized countries. Industrialized countries, on the other hand, are experiencing a certain saturation of energy service demand in most sectors. If C/gdp reduction targets were uniformly applied to developing and industrialized countries, developing countries would therefore still be put at a disadvantage.[3]

Per Capita Models

The Annual Per Capita Model. A recurring theme in the international debate over global equity has been the issue of population discrepancies between the developing and industrialized regions.[4] Proposals have been made to base emission reduction requirements on some annual per capita release rate that would eventually have to be achieved by all countries (see, e.g., Feiveson et al. 1988).

Compared to the carbon-release- or C/gdp-based models, per capita formulations allow a more rigorous equity approach: all human beings should be equally entitled to make use of global resources such as the planet's atmosphere. If some end up using more of such resources than others, they should provide some form of compensation (see also below and Chapter 7).

---

[3]In principle, this disadvantage could be remedied by establishing more lenient C/gdp reduction schedules for developing countries. Unfortunately, current analyses on the contribution of structural change to changes in the energy/gdp and C/gdp ratios are far too few and limited in scope to provide guidelines for such a differentiation. Also, developing countries may end up pursuing development models that differ significantly from the patterns found in industrializing nations in the past.

[4]See, for example, the summary of the population control debate in Eckholm (1982).

Of course, per capita formulations have their own drawbacks. First, the annual per capita model does not provide the incentive for population control inherent in the previous models: as formulated, the target per capita level would apply irrespective of the degree of population growth between now and the target year. Second, per capita models require the use of population forecasts to establish a link to the global carbon budget. This is less of a handicap than having to rely on gdp projections, since population forecasts are inherently more stable than gdp forecasts (see Section C). However, the use of projections could still prove awkward in international agreements.

Approaches based on an annual per capita target also fall short in addressing the dimension of *intergenerational equity*. This aspect is especially important in climate stabilization because it is a long-term task extending over many decades. How should the damage done by past generations be treated, and how can the interests of unborn future generations be explicitly incorporated into an equity formula?

The Cumulative Per Capita Model. As a modification of the per capita model, one could represent population changes over time (based on historic data or forecasts) in the form of "cumulative populations" for the period during which the phaseout of fossil fuels is to be achieved. The cumulative population over a period of time is simply the number of person-years lived in each region. For instance, a country with a population of 1 million in year x is counted as 1 million person-years.

Cumulative person-years would be calculated, in part, on the basis of population forecasts. Since the composition of present populations will largely determine future population levels over several decades, such forecasts are reasonably stable. The global carbon budget would then simply be allocated in proportion to the person-years lived in each region.[5]

Note that per capita models make no distinction between countries with an energy-intensive industrial mix and those whose economies are less energy-intensive (and therefore fossil-carbon-intensive). All else being equal, they favor nations that have managed to limit fossil fuel consumption in the past, and nations that are at early stages of

---

[5]The proposal by Feiveson et al. (1988) goes in the direction of controlling cumulative releases. It would establish a common long-term per capita target for all countries; put a short-to-medium-term ceiling on IC releases at current per capita levels, and would limit the degree to which DC per capita emissions could exceed the long-term target at any one time between now and the target date. These qualifications and specifications could be brought into approximate equivalence with the cumulative carbon budget.

economic development. Reduction requirements would appear relatively less drastic to them.

Summary. As a basis for sharing the global fossil carbon budget between developing countries and industrialized countries, the three models compare as follows:

- Under all formulas, major capital and technology transfer to the developing countries will be necessary to enable these countries to participate in climate stabilization. An equitable allocation of the global carbon budget can be seen as one of several instruments in this overall task.
- The lowest level of equity would be achieved in allocations of the cumulative fossil carbon budget on the basis of current release patterns.
- Uniform reduction milestones for C/gdp ratios would provide somewhat greater international equity, but would introduce target volatility and exacerbate measurement problems.
- The highest level of equity would be provided by allocating the fossil carbon budget on the basis of the cumulative per capita model. The cumulative person-year formula explicitly takes into account intergenerational dynamics while preserving the incentive for population control. It also could be used to develop equity formulas based on cumulative historic emissions (see below).

But is this strict equity model practical on logistic grounds, given the remaining allowable fossil fuel consumption and the inertia of existing consumption patterns? And if not, what might be a reasonable compromise?

These questions are explored in Sections C and D below. Section C illustrates the difference among the carbon-release-based and population-based models in a quantitative way, by applying them to the historic fossil carbon emissions between 1950 and 1986. In Section D, the analysis is extended to the future, covering the entire period between 1950 and 2100.

C. FOSSIL CARBON EMISSIONS AND INTERREGIONAL EQUITY: THE RECORD TO DATE

To illustrate the quantitative difference between the carbon-release-based and per capita models, we apply them to historic fossil carbon emissions as compiled by Marland et al. (1988). The carbon release dynamics between 1950 and 1986 are illustrated in Figures 5.4–5.9. In these figures, the world is alternately divided into two regions (developing countries [DCs] and industrialized countries [ICs]) or six regions: OECD Europe, North America (Canada and the United States), the Pacific (Australia, New Zealand, and Japan), Eastern Europe (including the Soviet Union), centrally planned Asia (China, Mongolia, Vietnam, North Korea), and all other developing countries.[6]

To put these figures in context, it is useful to first examine the current structure of world energy use.

Current World Energy Use and Fossil Fuel Consumption. Table 5.1 shows how the use of fossil fuels was distributed between industrialized and developing countries in 1986.[7] About 75 percent of the world's energy use was fossil-based, 15 percent was derived from biomass, 4 percent from nuclear power, and 6 percent from hydro power. Among fossil fuels, oil was dominant at 42 percent, followed by coal at 34 percent, and gas at 23 percent.

The industrialized nations accounted for three-quarters of fossil fuel consumption in 1986. On a per capita basis, the average person in the industrialized countries consumes 11 times as much fossil-based energy as the average person in the developing countries. If biomass fuels are included, the ratio is still 7:1.

Note also that fossil fuels account for only a little more than half of energy use in developing countries, whereas the fraction is 87 percent for the industrialized world. Coal is much more dominant in developing countries, while gas is correspondingly less important.

Annual Carbon Dioxide Emission Rates, Absolute. Figure 5.4 shows the development of annual carbon dioxide emissions from fossil fuel use between 1950 and 1986 for developing and industrialized countries. Figure 5.5 reproduces the same data disaggregated into six world regions. The ratio of releases in developing versus industrialized regions in 1986 was 1:2.6.

In 1950, the developing countries accounted for only 7.5 percent of total releases. By comparison, the share of Western Europe was

---

[6]The definition of regions and the fossil carbon emission data are from Marland et al. (1988). We consider only the 1950–1986 period, since regionalized carbon release data for the pre–World War II period are much less certain. Note that this procedure works in favor of industrialized countries.

[7]Fossil fuel consumption is based on U.N. statistics and higher heating values. Estimates of biomass consumption are based on per capita budgets developed by Goldemberg et al. (1988), and adjusted to reflect 1986 population levels.

TABLE 5.1   Approximate Global Distribution of Energy Use and Population in 1986[a, b]

	ICs	DCs	World		ICs	DCs	World
Population				**Biomass**			
(billion)	1.18	3.76	4.94	Total TW-y/y	0.07	1.80	1.88
share	*0.24*	*0.76*	*1.00*	*share*	*0.04*	*0.96*	*1.00*
				Fraction of total energy	*0.01*	*0.43*	*0.15*
Energy use				Per capita kW-y/y	0.06	0.48	0.38
Total TW-y/y	8.26	4.16	12.43	**Hydro**			
share	*0.66*	*0.33*	*1.00*	TW-y/y	0.47	0.23	0.70
Per capita kW-y/y	7.00	1.11	2.52	*share*	*0.67*	*0.33*	*1.00*
				Fraction of total energy	*0.06*	*0.06*	*0.06*
Fossil fuels				per capita kW-y/y	0.40	0.06	0.14
Total TW-y/y	7.21	2.11	9.32	**Nuclear**			
share	*0.77*	*0.23*	*1.00*	TW-y/y	0.51	0.02	0.53
Fraction of total energy	*0.87*	*0.51*	*0.75*	*share*	*0.96*	*0.04*	*1.00*
Per capita kW-y/y	6.11	0.56	1.89	*Fraction of total energy*	*0.06*	*0.00*	*0.04*
Fossil fuel Breakdown				per capita kW-y/y	0.43	0.01	0.11
Coal	*2.20*	*1.00*	*3.20*				
Oil	*3.04*	*0.89*	*3.93*				
Natural Gas	*1.97*	*0.22*	*2.19*	**Total non-fossil**			
				TW-y/y	1.05	2.05	3.11
Fraction of Total	*1.00*	*1.00*	*1.00*	*share*	*0.34*	*0.66*	*1.00*
Coal	*0.31*	*0.47*	*0.34*	*Fraction of total energy*	*0.13*	*0.49*	*0.25*
Oil	*0.42*	*0.42*	*0.42*				
Natural Gas	*0.27*	*0.10*	*0.23*	per capita kW-y/y	0.89	0.55	0.63

[a]Based on U.N. population and commercial energy statistics and biomass per capita estimates by Goldemberg et al. (1988).
[b]TW-yr/yr = Terawatt-year per year; kW = kilowatt-year per year.

24 percent. By 1986, the developing country share had climbed to 28 percent, while that of Western Europe was only 15 percent.

The historic data also show that until the 1973–1974 oil crisis, the United States was the largest emitter of CO_2 in absolute terms, with a growth rate in the 1960s that was unmatched by any other region. Second in growth and absolute numbers was the Comecon region (Soviet Union and Eastern Europe). Unlike in the United States and Western Europe, the Comecon releases did not drop after the second OPEC crisis in 1979–1980.

Today, releases from Eastern Europe and the Soviet Union match the U.S. emission rate. Meanwhile, emissions from developing countries experienced a rapid growth in the 1970s, fueled, in part, by the recent period of high economic growth in China (Chandler 1988; see also Chapter 6).

Cumulative Releases. The degree to which industrialized countries have inequitably consumed nonrenewable climatic resources is more

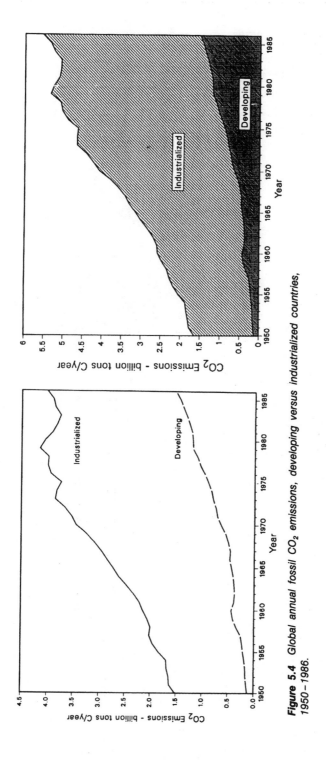

Figure 5.4 Global annual fossil CO_2 emissions, developing versus industrialized countries, 1950–1986.

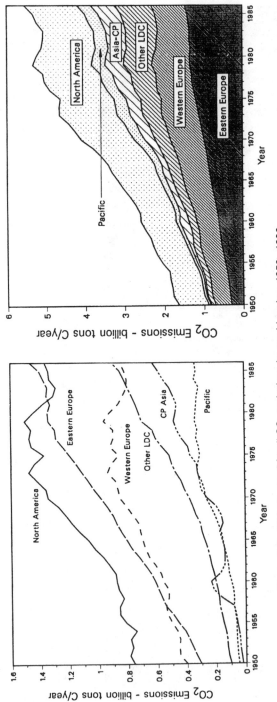

Figure 5.5 Global annual fossil CO_2 emissions by world region, 1950–1986.

226

accurately measured by the cumulative fossil carbon dioxide emissions. Due to lack of reliable data for the period before World War II, our comparison is limited to the period from 1950 to 1986. This period represents the era of greatest growth in energy consumption, and it captures the majority of cumulative fossil CO_2 releases to date (Rotty and Masters 1985).

The cumulative comparison among the regions in shown in Figure 5.6 (see also Table 5.4): only 18 percent of cumulative 1950–1986 emissions came from developing countries, compared to the 82 percent share of the industrialized countries. The ratio is thus 1:4.4, compared to the ratio of 1:2.6 for annual emissions in 1986.

This wider ratio still understates the real disparity, since a major portion of world cumulative releases occurred before 1950, and these were even more dominated by industrialized countries.

Per Capita Emission Rates. Figures 5.7 and 5.8 show the changes in per capita emission rates. Based on these figures, the gap between developing and industrial countries is even wider. In 1950, the average per capita fossil carbon dioxide releases in developing countries were only 4.1 percent of those in the industrialized countries, a ratio of 1:25. By 1986, the figure rose to 12.9 percent, or to a ratio of 1:7.8.

These numbers illustrate that the rise of developing nations into the league of major fossil carbon emitters (i.e., their 28 percent 1986

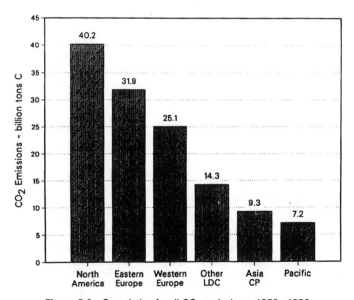

Figure 5.6 *Cumulative fossil CO_2 emissions, 1950–1986.*

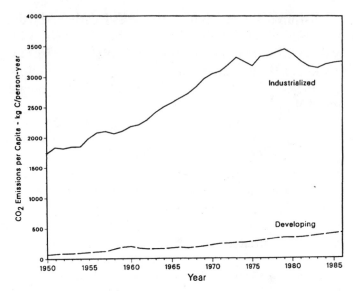

Figure 5.7   *IC and DC annual fossil CO_2 emissions per capita, 1950–1986.*

annual contribution, or their 19 percent 1950–1986 cumulative share) is much less true on a per capita basis. Per capita, industrialized countries still release about eight times the amount of fossil CO_2 emitted by Third World countries.

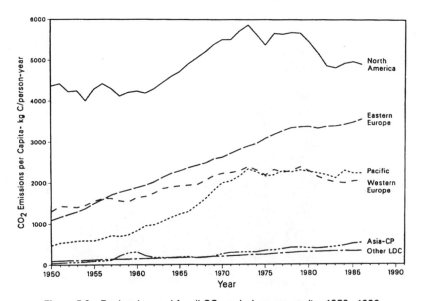

Figure 5.8   *Regional annual fossil CO_2 emissions per capita, 1950–1986.*

Cumulative Per Capita Emissions. The most accurate measure of the relative consumption of climatic resources in industrialized and developing countries is obtained by correlating cumulative releases with the development of the populations in the two regions. We show this comparison in Figure 5.9, based on U.N. population statistics: between 1950 and 1986, industrialized countries released 11.1 times as much fossil carbon dioxide per person-year as developing countries. Thus, on this account, the imbalance between the two groups of countries widens even further.

D. EQUITY-BASED ALLOCATION OF THE GLOBAL CARBON BUDGET

Below, we calculate the regional allotment of fossil carbon emissions on the basis of *cumulative per capita emissions*. This approach is applied to two periods: 1950–2100 and 1986–2100. The former accounting period amounts to treating historic carbon releases between 1950 and 1985 as a "cash advance" against the total 1950–2100 carbon budget. The latter time frame ignores historic inequities in fossil fuel consumption. Instead, past fossil carbon releases are "grandfathered."

In extending the cumulative per capita approach to future time

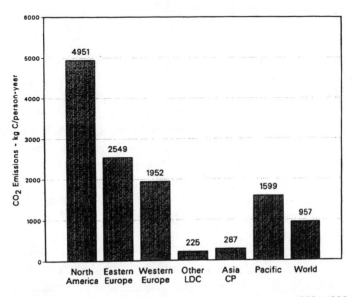

Figure 5.9 *Cumulative fossil CO_2 emissions per capita-year, 1950–1986.*

periods, a population forecast has to be used. As mentioned above, this is the major drawback of this model when considered as a basis for international agreements. A simple way around this problem would be to apply historic cumulative populations. To examine the viability of this approximation, we compare the allocation shares based on the 1950–1985 period with those derived from long-term population forecasts.

1. Population Forecasts and Cumulative Person-Years

Population growth is in some sense like global warming: due to the inertia of intergenerational dynamics, a large amount of population growth is already unavoidable though not yet realized. For this reason, long-term population forecasts are relatively more reliable than other kinds of forecasts. Changes in future birthrates, life expectancies, and other factors such as urbanization occur slowly and take several decades to show up in significantly altered population trends.

A widely used long-term population forecast is that prepared by the United Nations (U.N. 1982; 1984). Table 5.2 summarizes the United Nations' high, medium, and low projections. The range of projections for 2100 is anywhere from about 7.5 to 14.2 billion.

A key parameter for understanding these scenarios is the year in which a net reproduction rate of 1.0 is reached. This is the replacement level fertility that would eventually stabilize populations. In the U.N. low projection, this point is reached as early as 2010. In the high projection, it is not reached until 2065, and in the medium case, it is reached by 2035.

Current population control programs are plainly insufficient to achieve replacement levels by 2010, or for that matter, by 2035. Equally important, prospects are extremely poor that Third World standards of living will soon broadly rise to levels where high birthrates would truncate on account of income effects. In view of this constraint, the U.N. low projection may no longer be attainable in an orderly fashion. The most likely but horrifying scenario leading to low population figures is that catastrophic deterioration in the world's environmental conditions would drastically shorten life expectancy. From a status quo perspective, the high forecast would appear to be the most likely scenario.

On the other hand, if the world proceeds along the high scenario, population pressures on natural resources could become so intense that the earth's carrying capacity would be exceeded. Rather than leading to a population of up to 14 billion people in 2100, as projected

TABLE 5.2   Population Forecasts (in millions) for the World and by Regions, 1985–2100

	Developing Countries			Industrialized Countries			World		
	Low	Medium	High	Low	Medium	High	Low	Medium	High
1980	3290	3301	3308	1130	1131	1132	4420	4432	4440
1985	3578	3634	3674	1155	1165	1173	4733	4798	4847
1990	3892	4000	4080	1180	1199	1215	5072	5199	5295
1995	4233	4403	4532	1206	1235	1259	5439	5638	5790
2000	4604	4847	5033	1233	1272	1304	5837	6119	6337
2005	4841	5189	5472	1237	1292	1339	6078	6482	6811
2010	5090	5556	5950	1240	1313	1375	6330	6869	7324
2015	5352	5948	6469	1244	1334	1411	6596	7282	7880
2020	5627	6368	7033	1247	1355	1449	6875	7724	8483
2025	5917	6818	7647	1251	1377	1488	7168	8195	9135
2030	6028	7059	8071	1239	1382	1512	7267	8441	9583
2035	6141	7308	8519	1228	1387	1536	7369	8695	10055
2040	6256	7567	8992	1216	1392	1560	7473	8959	10552
2045	6373	7834	9491	1205	1397	1585	7579	9231	11076
2050	6493	8111	10018	1194	1402	1610	7687	9513	11628
2055	6497	8221	10326	1185	1405	1628	7682	9627	11954
2060	6500	8333	10643	1177	1409	1646	7677	9742	12289
2065	6504	8446	10970	1168	1412	1664	7672	9858	12634
2070	6507	8561	11307	1159	1416	1682	7667	9976	12989
2075	6511	8677	11654	1151	1419	1701	7662	10096	13355
2080	6486	8694	11812	1148	1419	1707	7634	10114	13519
2085	6461	8712	11972	1145	1420	1714	7606	10131	13686
2090	6436	8729	12135	1143	1420	1720	7579	10149	13855
2095	6412	8747	12299	1140	1421	1727	7551	10167	14026
2100	6387	8764	12466	1137	1421	1733	7524	10185	14199

Source: U.N. (1982; 1984).

under conditions of trend continuity, this scenario could, for example, lead to significantly lower population levels—but again due to catastrophic breakdowns. Nor is the U.N. medium scenario immune to such an outcome.

For the purposes of the study in this book, we neglect such feedback risks. We further assume that a successful international cooperation effort to curb climate warming would strengthen population control programs and thus lower world population growth. Success in implementing broad policies for sustainable development should also lead to substantial income gains in the developing countries. We therefore use the U.N. medium forecast in future discussions.

Regional Distribution of Person-Years: Past and Future. Table 5.3 shows the distribution of person-years between developing and industrialized countries. In the period from 1950 to 1986, developing countries accounted for about 95 billion person-years, or 71 percent of the

TABLE 5.3   Population Dynamics and Cumulative Person-Years, 1950–2100

Year	Population (billion) [1]			Person-years (billion) [2]		
	ICs	DCs	World	ICs	DCs	World
1950	0.77	1.75	2.52			
				38	95	133
1986	1.18	3.76	4.94			
				17	60	77
2000	1.27	4.85	6.12			
				33	145	178
2025	1.38	6.82	8.20			
				35	186	221
2050	1.40	8.11	9.51			
				35	210	245
2075	1.42	8.68	10.1			
				36	218	254
2100	1.42	8.76	10.2			
1950–1986 share				38	95	133
				0.29	*0.71*	*1.00*
1986–2100 share				156	819	975
				0.16	*0.84*	*1.00*
1950–2100 share				194	914	1108
				0.18	*0.82*	*1.00*

[1]Population projections based on U.N. medium case (U.N. 1982; 1984).
[2]Approximate values based on interpolation.

total. This share is not much different from the 1986 Third World share of world population (76 percent) or its historic share between 1950 and 1985 (74 percent). Between 1986 and 2100, the DC share would be 84 percent of the global person-year total. If the entire period from 1950 to 2100 is taken as a basis, the DC share becomes 82 percent.

These figures show that the person-year percentages are not very sensitive to changes in the base period. To reduce the contentiousness of allocation shares in international negotiations, equity calculations could either be based on shorter population forecast periods (e.g., the next 50 years, in which the low-to-high forecast variation is only 36 percent compared to 89 percent in 2100, [see Table 5.3]); or they could simply be based on historic figures.

2. Cumulative Per Capita Equity: Results

Regional Fossil Carbon Budgets: Allocation Basis 1950–2100.
Under a 400-ppm/300-btC ceiling for the 1986–2100 period, the total fossil carbon budget from 1950 to 2100 would be 428 btC (see Table

5.4). According to Table 5.3, the person-year-based share of industrialized countries would be 18 percent of the overall 1950–2100 budget, or 77 btC. In actuality, industrialized countries released 105 btC between 1950 and 1986. This result shows that when historic emissions are counted, industrialized countries have already overdrawn their entire 1950–2100 account. This finding is not altered if, for the sake of sensitivity analysis, a 450-btC budget and/or the U.N. low-population forecast is assumed.

For the developing nations, the converse is true: their 1950–2100 fair share of 351 btC has been greatly "underused" to date. If this much emission were still permissible, it would last them for another 255 years at the current Third World fossil fuel consumption rate and mix (Table 5.4).

Regional Fossil Carbon Budgets: Allocation Basis 1986–2100. If we ignore the past and just divide the 300 btC (1986–2100) carbon budget among the generations living between 1986 and 2100, things don't look much better: according to Table 5.3, the fair share of industrialized countries would be 16 percent, or 48 btC. At current rates of fossil fuel consumption, this budget would last about 12 years

TABLE 5.4   Historic and Remaining Permissible Fossil CO_2 Releases, Cumulative Per Capita Equity under a 400-ppm/300-btC Ceiling, 1950–2100 and 1985–2100

	World billion t C	ICs billion t C	DCs billion t C	Years left @ 1986 release rate		
				ICs	DCs	World
Actual (1)						
1950	1.55	1.44	0.11			
share	1.00	0.93	0.07			
1986	5.27	3.90	1.37			
share	1.00	0.74	0.26			
1950–1986	127.9	104.4	23.5			
share	1.00	0.82	0.18			
Equity basis						
1986–2100	300	48	252	12	183	57
share (2)	1.00	0.16	0.84			
Equity basis						
1950–2100	428	77	351	—	255	81
share (2)	1.00	0.18	0.82			

(1)Historic data from Marland et al. (1988), including cement, without bunkers.
(2)Share of 1986–2100 cumulative person-years, based on Table 5.3.

(Table 5.4). Even under a 450-btC budget, and assuming the U.N. low-population forecast, the total would not last longer than 20 years.

By contrast, the fair share of the 300-btC budget for developing countries would allow them to operate for 183 years at their current level of consumption.

Conclusions. The above information suggests the following conclusions:

- The intensely fossil-based growth of industrialized nations has forever preempted Third World nations from using fossil fuels to a similar extent for their own development.
- The high consumption of fossil fuels in industrialized countries has made a strictly equitable sharing of the remaining permissible fossil carbon budget impossible. The fair carbon budget share of industrialized countries would not be sufficient to allow them an orderly phaseout of fossil fuels.
- If climate stabilization is to be achieved, developing countries will need to leapfrog into economic development based on nonfossil energy sources.

3. What Amount of Fossil Fuels Is Needed for Development?

There is an important irony to this situation which has not been widely recognized so far. Could developing countries make productive use of their share of the global carbon budget if it were indeed still available? The answer is probably no. The mere fact that remaining allowable global carbon emissions are so limited means that any economic infrastructures built up mainly on the basis of fossil fuels risk early obsolescence. In effect, the tight carbon budgets implied by climate stabilization greatly reduce the long-term value of fossil fuels to developing countries:

- If any policy of global climate stabilization is pursued at all, industrialized countries will be making rapidly increasing use of existing and new high-efficiency technologies to reduce their carbon releases. Such technologies tend to promote energy efficiency and productivity advantages in tandem.
- Without use of these fossil-substituting technologies in their own modern sectors, developing countries would risk losing competitiveness in international markets. In that sense, overly heavy reli-

ance on fossil fuels during periods of intensive infrastructure buildup would create inefficiencies, with a serious retarding effect on development in later years.[8]

To be sure, increased access to fossil fuels is necessary if developing countries are to overcome their near- and medium-term economic difficulties. But increased reliance on fossil fuels in the near and medium term will only be developmentally beneficial for them if it is part of a broader strategy of permanent fossil fuel displacement.

If these issues are taken into serious consideration, Third World countries might want to use much less fossil fuel than represented by the roughly 250 btC they would be theoretically entitled to under our strict cumulative per capita criterion.

To summarize,

- Fossil fuel consumption in the developing countries based on a cumulative per capita allocation of the fossil carbon budget is not only infeasible from a logistic point of view but appears also undesirable from a developmental point of view.

However, the cumulative per capita criterion could still be used in the service of international equity. Such options are explored in the concluding section of this chapter.

E. PERSON-YEAR EQUITY AND COMPENSATORY INTERNATIONAL ASSISTANCE

While the cumulative person-year model is not practical for the *physical* distribution of remaining permissible fossil carbon emissions, it could be used to allocate the *financial* burden of supporting Third World energy development. But to arrive at an acceptable formula, it will be desirable to eliminate population projections, which are liable to be a source of contention in international negotiations. Two basic approaches are worth discussing:

1. Developing countries could be given emission rights based on their share in cumulative person-years between 1950 and the present. As suggested by our analysis in Sections C and D, this

[8]See footnote 3.

approach would give them emission rights in excess of their short- and medium-term needs, if not of their long-term needs as well. They could thus sell emission rights to industrialized countries, who would be short of such rights. The proceeds would be used to boost energy efficiency, reforest and conserve soils, and finance the development of nonfossil energy systems.

2. Alternatively, industrialized countries could pay into a "climate protection fund" which would be used for capital and technology transfer programs in proportion to some equity formula, for example, their cumulative per capita releases between 1950 and the present. Again, the proceeds would be used to finance proposals by developing countries for boosting energy efficiency, reforesting and conserving soils, and financing the development of nonfossil energy systems.

Both arrangements could provide the necessary flexibility in physically sharing the global carbon budget while tying financial compensation to strict equity criteria. Industrialized countries could raise payments for assistance programs or for purchasing emission rights by levying taxes on energy use in general, and fossil fuel use in particular.

Some of these IC revenues could also finance the accelerated development of low-cost efficiency and renewables-based technologies at home. Within a few decades, when income growth in developing countries would lead to major increases in energy service demand, affordable nonfossil technologies would be widely available and could allow developing countries to complete their leapfrog into economies based on highly efficient use of modern renewables.

1. Tradable Emission Rights

Tradable emission rights have long been advocated as an economically efficient means of implementing environmental goals.[9] The debate over global warming has rekindled interest in the concept (see e.g., von Weizsaecker 1988). Emission rights are directly compatible with the concept of cumulative emission budgets developed in this study, since the definition of emission rights depends on an environmental target level or ceiling for pollutant emissions.

Theoretically, tradable emission rights could offer an effective means of providing developing countries compensation for lost oppor-

[9]For an early discussion of this and other forms of rationing environment impacts, see, for instance, Westman (1976).

tunities to use fossil fuels. If workable, they could lead to a sharing of the global carbon budget based on market efficiency. Rigid allocation formulas that are difficult to adapt to changing needs and circumstances would be avoided.

Note that our sketch deviates in a major way from the usual emission rights proposals made in the context of conventional air pollution policy: base-period emission rights are not awarded in proportion to existing emission patterns, but according to a formula for intergenerational per capita (i.e., person-year) equity. Unlike conventional proposals, this feature would force the market to start working from an equitable baseline distribution of emission rights rather than from a "grandfathered" status quo.

Limitations and Problems. Whether tradable emission rights can serve as a central allocation and compensation instrument is questionable. A number of potential implementation problems are readily apparent:

- Tradable emission rights reduce the complexities of each individual carbon-substituting investment to a monetary value. However, many important aspects of such investments, including impacts on local communities, national self-reliance, and environmental effects cannot be adequately captured by market prices based solely on carbon benefits.
- A purely market-based approach could conflict with the long-term planning that is required to observe the limits of the global fossil carbon budget. The year-to-year flows of emission rights from poor nations to rich nations and vice versa could create uncertainty that might interfere with monitoring of timely plan implementation and other long-term planning activities.
- Under the burden of tremendous debt and other short-term difficulties, governments of developing countries could be forced to sell large amounts of certificates at a time (the bulk of which they need only in the longer term), sending prices to the fire-sale level. This would undermine the compensation payment and capital and technology transfer objectives of the arrangement. Alternatively, developing countries might be forced to form an "emission rights OPEC."
- The same short-term pressures, and/or exaggerated hopes for technology innovations, might lead governments of capital-poor countries to sell too much of their emission rights early on. In later

years, they may find themselves unable to live within their remaining carbon budgets, but may not have the capital or the opportunity to buy them back.

- All this could undermine the equity objective of the initial assignment of emission rights: the rich nations might end up buying their way out of painful adjustments, while the poor nations could not.
- The dominance of the energy supply industries of the industrial nations over energy planning in the developing countries could be reinforced, which could undermine efforts by developing nations to achieve technological and economic self-reliance (see also Chapter 7).

The development experience to date, and specifically the Third World debt crisis, suggests that approaches based mainly on market exchange will not work among nations as structurally different as developing and industrialized countries (Senghaas 1977; 1982). Such approaches will tend to reinforce existing relationships of economic dominance—the same relationships that already are a major cause of environmental destruction. Rather than allowing developing countries to be part of the solution, this self-feeding dynamic could ultimately make them less capable of contributing to climate stabilization.

Of course, the emission rights approach could probably be modified to eliminate a number of its drawbacks. For example, the amount of fossil carbon release rights that would be tradable in any year could be limited to a small percentage of each country's or region's total, or to a fraction of annual increments derived from long-term reduction targets. Trade could further be limited to industrialized countries, or organized within trading groups that are reasonably matched in terms of levels of economic development. Such rules could limit price drops and support long-term planning.

But if trading is limited in these ways, it may not provide a sufficient mechanism for generating the large capital transfers that will be needed to get Third World investments under way. Emission rights trading could likely be no more than one of several elements in an overall redistribution/fossil carbon reduction strategy.

2. A Climate Protection Fund Combined with Carbon Reduction Auctions

The idea of a climate protection fund has appeared in a number of proposals for a global climate convention (see Chapter 7). The task of the fund would be to support implementation of long-term DC and IC regional reduction goals (see Chapter 6) as negotiated in the climate convention protocols (see Chapter 7). A principal use of the fund in the area of fossil carbon reductions would be to pay capital-strapped developing countries for the extra first cost of installing and/or retrofitting the most energy- and pollutant-efficient end-use technologies instead of standard technology, and for the extra cost of renewables-based supply systems.

Such an international fund might function as follows:

- *Initial Capitalization.* Initial contributions would come from industrialized countries and would be in proportion to the cumulative or cumulative per capita emissions between 1950 and the present. This start-up capitalization would be similar to the Super Fund approach used, for example, in the United States for funding cleanups of toxic waste sites. In this case, the atmosphere is the toxic waste site. Table 5.5 shows each region's payment for a start-up capitalization of about $100 billion, which could be mobilized over a number of years. As noted in Section B, the two formulas are different in their impact on regions. Obviously, ability to pay must be taken into account in the case of Eastern Europe.

- *Ongoing Contributions.* Ongoing contributions could be tied, wholly or in part, to actual releases or to progress in reducing releases according to some schedule. Like the trading of emission

TABLE 5.5 Illustration of Possible Equity Formulas for the Initial Capitalization of a Global Climate Protection Fund

Region	Cumulative emission equity			Cumulative per capita emission equity [2]		
	Cumulative fossil C releases 1950-86 [1] btC	Share of IC total	Initial Fund capitalization @ $1/ton C in 1, billion U.S.$	Cumulative per capita emissions ton C/pers.-yr	Index	Initial Fund capitalization of $104 billion based on 4,5 billion U.S.$
	1	2	3	4	5	6
North America	40.2	0.39	40.2	4.95	1.79	46.8
Western Europe	25.1	0.24	25.1	2.55	0.92	24.1
Eastern Europe	31.9	0.31	31.9	1.95	0.71	18.4
Pacific/Other	7.2	0.07	7.2	1.60	0.58	15.1
All ICs	104.4	1.00	104.4	2.76	1.00	104.4

[1] Historic data from Marland et al. (1988), including cement, excluding bunkers.
[2] Definition of cumulative per capita emissions as in Section B.

rights, this would give an economic incentive for each nation to reduce carbon releases over time.

For example, ongoing payments could be based on the difference between actual releases in a given year and average annual releases based on a 1986–2100 cumulative IC carbon budget. Table 5.6 shows such a calculation for an IC cumulative budget of 150 btC for the principal industrialized regions.[10] For purposes of illustration, the table also shows the annual capital contributions based on a $15/t carbon fee (about $0.001/kWh primary fossil energy at the current fuel mix, or equivalent to about 10 percent of 1986–1990 fossil energy prices). This would raise some $40 billion annually, comparable to the funds now available to bilateral and multilateral agencies. Under this formula, Portugal is the only nation for which fossil carbon emissions are below the threshold for becoming a fund contributor. Again, the economic conditions in each country will need to be taken into account. In particular, the countries of Eastern Europe might find it difficult to make large annual contributions at this time.

- *Solicitation of Carbon Reduction Proposals.*   An international oversight agency would determine each year what decrement of fossil carbon reductions is needed in each world region. The decrement would be calculated on the basis of agreed-upon long-term reduction schedules, and monitoring results from previously funded projects. The fund would then issue a request for proposals, and developing countries would file concrete project proposals with specified carbon benefits.
- *Mechanisms and Criteria for Project Selection.*   The annual reduction increment in each region would be filled by the proposals with the lowest cost and greatest nonmonetary benefits. An auction or bidding mechanism[11] could be used to help determine the

---

[10]Cumulative budgets would be converted into average annual release rates by dividing them by the number of years in the cumulation period. To illustrate this formula, take Western Europe, which had a 31 percent share in 1986 IC population (Table 5.6). Under a 150-btC cumulative IC carbon budget, this translates into emission rights of 46.6 btC, or an average of 46.6/115 = 0.405 btC/yr. Current annual emissions are about 0.87 btC. Payments into the fund would be proportional to the difference of 0.87 − 0.405 btC/yr, which is 0.47 btC/yr. As Western Europe reduces its fossil carbon releases over the years, payments would become proportionately smaller.

[11]A model for such bidding, as well as a valuable source of experience with the use of price and nonprice factors, is provided by the utility resource auctions that have recently been implemented in several states of the United States.

TABLE 5.6 Illustration of a Possible Equity Formula for Funds Contributions from ICs

Region or country	Population in 1986 (millions)	Share of IC total	Reference 1986-2100 fossil C-budget [1] btC	Reference ave. permissible releases [2] btC/year 4	1986 releases [3] btC/year 5	Contribution Index btC/year (5) – (4)	Annual Contribution @$15/ton C [4] U.S. $ billion
North America	267	0.220	32.9	0.286	1.22	0.93	14
Western Europe	378	0.311	46.6	0.405	0.87	0.47	7
Eastern Europe	393	0.323	48.5	0.422	1.40	0.98	15
Pacific	141	0.116	17.4	0.151	0.32	0.17	3
Other [5]	37	0.030	4.6	0.040	0.10	0.06	1
All ICs	1216	1.000	150.0	1.304	3.92	2.61	39
Australia	16	0.013	2.0	0.017	0.061	0.044	0.66
Austria	7.6	0.006	0.9	0.008	0.014	0.006	0.09
Bulgaria	9	0.007	1.1	0.010	0.033	0.023	0.35
Canada	25.6	0.021	3.2	0.028	0.105	0.078	1.17
Czechoslovakia	15.5	0.013	1.9	0.017	0.066	0.049	0.74
Denmark	5.1	0.004	0.6	0.005	0.017	0.012	0.17
FR Germany	60.9	0.050	7.5	0.065	0.186	0.121	1.81
Finland	4.9	0.004	0.6	0.005	0.015	0.010	0.14
France	55.4	0.046	6.8	0.059	0.098	0.039	0.58
Germ. Dem. Rep.	16.6	0.014	2.1	0.018	0.092	0.074	1.12
Greece	10	0.008	1.2	0.011	0.016	0.005	0.08
Hungary	10.6	0.009	1.3	0.011	0.021	0.010	0.15
Iceland	0.2	0.000	0.0	0.000	0.000	0.000	0.00
Ireland	3.6	0.003	0.5	0.004	0.008	0.004	0.06
Israel	4.3	0.004	0.5	0.005	0.007	0.003	0.04
Italy	57.2	0.047	7.1	0.061	0.095	0.033	0.50
Japan	121.5	0.100	15.0	0.130	0.256	0.126	1.89
Luxemburg	0.4	0.000	0.0	0.000	0.002	0.002	0.03
Netherlands	14.6	0.012	1.8	0.016	0.035	0.019	0.29
New Zealand	3.3	0.003	0.4	0.004	0.005	0.002	0.03
Norway	4.2	0.004	0.5	0.005	0.009	0.004	0.06
Poland	37.5	0.031	4.6	0.040	0.124	0.084	1.26
Portugal	10.2	0.008	1.3	0.011	0.008	-0.003	-0.04
Romania	22.9	0.019	2.8	0.025	0.056	0.031	0.47
South Africa	32.3	0.027	4.0	0.035	0.093	0.058	0.87
Spain	38.7	0.032	4.8	0.041	0.050	0.008	0.13
Sweden	8.4	0.007	1.0	0.009	0.016	0.007	0.11
Switzerland	6.5	0.005	0.8	0.007	0.011	0.005	0.07
United Kingdom	56.7	0.047	7.0	0.061	0.166	0.105	1.58
United States	241.6	0.199	29.8	0.259	1.202	0.942	14.14
USSR	281.1	0.231	34.7	0.302	1.011	0.709	10.64
Yugoslavia	23.3	0.019	2.9	0.025	0.035	0.010	0.15

[1] Reference budget for contribution calculations based on 300-btC global budget, 50% IC share (see Chapter 6), allocated among ICs according to column 2.
[2] Reference average permissible release rate based on reference budget applied to 1986–2100 period.
[3] Regional release rates from Marland et al. (1988), including cement, excluding bunkers. These may vary from national data calculated from more detailed statistics or using different heating value conventions.
[4] $15/tC at the 1986 global fuel mix is equivalent to a roughly 10% increase of 1986 crude oil prices.
[5] Other is South Africa and Israel.

most attractive projects and to assure efficient use of capital. Proposals would be evaluated according to a point system that includes both price and nonprice factors. The nonprice factors
would emphasize ecodevelopment criteria (see Chapter 7).

Many other variants of the basic scheme could be conceived. Note that under the above formulas, the largest amounts of capital for DCs would be available in the initial two to four decades, when industrialized country emissions will be most out of line with their fair share, and the need for massive investments in developing countries will be most pressing.[12]

Of course, the bidding/climate fund approach has its own drawbacks. Given that much climate-stabilizing action will need to occur through dispersed local action, an international agency may have inherent difficulties controlling the use of funds without stifling creative, self-reliant approaches. A large, highly centralized bureaucracy could prove inefficient in administering such tasks (see also Section E in Chapter 3). Industrialized nations may balk at giving developing countries significant control over how available funds would be used. Agreement on scoring systems for the competitive bidding process could be difficult to reach. And in the area of modern sector energy investments, many developing countries may find themselves short of the necessary technical sophistication to gear their proposals to the most viable technological options.

3. The Need for Reduction Milestones

The above compensatory financial mechanisms could in principle ensure that poor countries can meet their development goals with less than their strict cumulative per capita equity share of the global carbon budget. But the adequacy of both the tradable emission rights approach and the climate-fund mechanism is predicated on the existence of equitable agreements about reduction milestones. Only once milestone agreements have been set in accordance with the global fossil carbon budget can these mechanisms reliably assist climate stabilization.

How could equitable reduction targets be determined? The use of compensatory financial mechanisms frees negotiations from the dilemma of the infeasibility of strict equity formulas. Nevertheless, basic equity considerations still apply: it is fair to argue that if industrialized

---

[12]The manner in which industrialized countries would implement their reduction targets at home is not addressed by this text. Price-based incentives such as carbon taxes are by themselves insufficient to spur the necessary reductions. Standards, financial incentives to consumers, and other policies will need to be implemented as well. Furthermore, surcharges should not be limited to fossil fuels alone, but should also be applied to other sources of energy, since all of them entail risks and environmental impacts.

countries cannot manage with their fair share of global fossil carbon budget, they should at least displace fossil fuels as fast as they can without economic disruption.

This course of action would limit the world's exposure to the risk of climate disruption to the unavoidable minimum. It would create the most rapid expansion of commercially available fossil-substituting technologies for all countries, while providing developing countries with the necessary leadership by example. With appropriate technology transfer, Third World nations could base most of their new capital stocks on modern nonfossil and high-efficiency technologies.

The key question for our further analysis thus becomes

- How fast could industrialized countries phase out fossil fuels without creating serious economic disruptions?

This question leads us to a broader issue that needs yet to be addressed in our overall analysis: Is a 300-btC cumulative fossil carbon budget an economically and logistically feasible ceiling? While economic and logistic factors are ultimately interrelated, a detailed exploration of the economic cost of phasing out fossil fuels is not included in this text. The overall rate and magnitude questions, and the equity issues implied in them, are explored in the following chapter.

REFERENCES

Chandler, W. U. (1988), "Assessing Carbon Emission Control Strategies: The Case of China," *Climatic Change*, Vol. 13, No. 3, pp. 241–265.

Eckholm, E. P. (1982), *Down to Earth*, International Institute for Environment and Development (IIED), Norton, New York.

Feiveson, H. A., et al. (1988), *Princeton Protocol on Factors That Contribute to Global Warming*, Woodrow Wilson School of Public and International Affairs, Princeton University, NJ.

Goldemberg, J., T. B. Johannson, A. K. N. Reddy, and R. H. Williams (1988), *Energy for a Sustainable World*, Wiley, New York.

Haefele, W., et al. (1981), *Energy in a Finite World*, Ballinger, Cambridge, MA.

Houghton, R. A., et al. (1987), "The Flux of Carbon from Terrestrial Ecosystems to the Atmosphere in 1980 Due to Changes in Land Use: Geographic Distribution of the Global Flux," *Tellus*, Vol. 39B, No. 1–2, pp. 122–139.

Krause, F. (1980), *Materialienband zur Energiewende*, Oeko-Institut, Freiburg/Darmstadt, FRG.

Lashof, D. A. and D. A. Tirpak, eds. (1989), *Policy Options for Stabilizing*

Global Climate, Draft report to the U.S. Congress, U.S. Environmental Protection Agency, Office of Policy, Planning, and Evaluation, Feb.

Lovins, A. B., L. H. Lovins, F. Krause, and W. Bach (1981), *Least-Cost Energy: Solving the CO_2 Problem,* Brick House, Andover, MA. (German version: *Wirschaftlichter Energieeinsatz: Losung des CO_2-Problems,* Karlsruhe, FRG, 1983.)

Marland, G., et al. (1988), *Estimates of CO_2 Emissions from Fossil Fuel Burning and Cement Manufacturing Using the United Nations Energy Statistics and the U.S. Bureau of Mines Cement Manufacturing Data,* Carbon Dioxide Information Analysis Center (CDIAC), Numeric Data Collection, Oak Ridge National Laboratory, Oak Ridge, TN.

Reddy, A. K. (1989), Personal communication, ASTRA, Indian Institute of Science, Bangalore, India.

Rotty, R. M., and C. D. Masters (1985), "Carbon Dioxide from Fossil Fuel Combustion: Trends, Resources, and Technology Implications," in J. R. Trabalka, ed., *Atmospheric Carbon Dioxide and the Global Carbon Cycle,* U.S. Department of Energy, Washington, D.C., pp. 63–80.

Senghaas, D. (1977), *The International Economic Order and the Politics of Development* (in German), Suhrkamp, Frankfurt.

Senghaas, D. (1982), *Lessons of the European Development Experience* (in German), Suhrkamp, Frankfurt.

U.N. (1982), *Population Bulletin of the United Nations,* Department of International Economic and Social Affairs, United Nations Secretariat, New York, No. 14.

U.N. (1984), *Population Bulletin of the United Nations,* Department of International Economic and Social Affairs, United Nations Secretariat, New York.

U.N. (1986), *Yearbook of World Energy Statistics,* United Nations, New York.

Westman, W. (1976), "Rationing Environmental Impacts Using a Second Currency," *Journal of Environmental Management,* Vol. 4, pp. 355–381.

von Weizaecker, C. F. V. (1988), Statement before the Enquête-Kommission Preventative Protection of the Earth's Atmosphere of the West German Parliament (Deutscher Bundestag, Enquête-Kommission Vorsorge zum Schutz der Erdatmosphaere), Bonn, Nov. 25–26.

6

How Quickly Must Fossil
Fuels Be Phased Out?
Global and
Regional Milestones

A. INTRODUCTION

Within a carbon budget framework, reduction targets could in principle vary over a wide range. In practice, that range will be narrowed by logistic constraints and considerations of political economy.

In this chapter, we investigate the degree of latitude in timing that is available to society in weaning itself from fossil fuels, given a 2–2.5°C mean global warming limit. As before, climatic necessity is represented in the form of the 300-btC cumulative fossil carbon budget for the period 1985–2100. To establish a practical band of reduction milestones, we estimate what phaseout rates for fossil fuel burning would be reasonable, given political economy and logistic constraints in restructuring the energy sector. We then examine whether these pragmatic rates of phaseout would be compatible with the global fossil carbon budget, and make modifications as needed. Phaseout schedules are thus arrived at in an iterative manner.

In all this, we apply our analysis of the global equity issue from Chapter 5. As a simple working hypothesis, the global fossil carbon budget is split evenly between industrialized and developing countries. This choice represents an approximate midpoint between the cumulative per capita formula and the carbon release formula. We examine the implications of this split for the required speed and practical feasibility of the fossil fuel phaseout in each group of countries, and develop regionalized phaseout schedules.

Specifically, we identify the approximate range of dates by which the world and the two regions will have to achieve a 20 percent, 50 percent, and 75 percent reduction in carbon releases. We also discuss various sensitivity cases, including a 450-btC budget.

To establish a connection to existing target proposals, the Toronto target (a 20 percent reduction in fossil carbon releases by 2005) is also examined. We calculate this target's compatibility with the global fossil carbon budget when applied to the world as a whole, and when applied to industrialized countries only, using the same equity, political economy, and logistic considerations.

Finally, the changes in *C*/gdp ratios implied in these trajectories are compared with historic patterns in selected countries and groups of countries.

The setting of phaseout targets will likely bring forth political pressures from the powerful commercial enterprises that produce, transform, and sell fossil fuels. There are other societal groups and forces whose immediate interests will be negatively affected by a fossil phaseout. Economic regions that are heavily dependent on energy production, and labor unions in these industries immediately come to mind. In several countries, the coal regions are important political bases for the major labor-oriented political parties. And the geopolitical significance of nations with large fossil reserves will also be strongly affected. The greenhouse issue will therefore fuel the politicization of energy policy. This means that factors of political economy must be carefully considered in constructing phaseout schedules.

At the same time, there are logistical constraints on the speed of phasing out fossil fuels. Principally, these are related to the sheer magnitude of capital stocks that need to be restructured. But an orderly fossil phaseout also requires that time be allowed for restructuring the many regulatory and institutional infrastructures that now surround the supply of energy services.

B. WHAT RATES OF FOSSIL FUEL PHASEOUT COULD BE VIABLE?

We begin with the political economy aspects of the problem, followed by a brief exploration of logistic constraints. The two aspects are, of course, interrelated. Political economy analyses usually examine how economic interests are intertwined with the institutions that administer regulatory decisions, and with regulatory approaches themselves. The focus is on a somewhat narrower issue here: the "economic pain thresholds" that are likely to shape the reaction of various interest groups to regulatory action, given the nature of their investment in the status quo pattern of fossil fuel use.

1. Least-Cost Efficiency versus Sunk Investments

In the following analysis, it is assumed that all affected parties will be willing to make some adjustments to help achieve climate stabilization. We further postulate that the most legitimate and most widely acceptable approach would be one that achieves the phaseout of fossil fuels at least cost. What mix of nonfossil or low-carbon resources could provide such a least-cost phaseout is explored in detail in Krause et al. (1991). Here, we are only concerned with the temporal aspect of a least-cost phaseout, that is, the effect of the speed of fossil carbon reductions on the cost of these reductions. For this discussion, we must clarify the major perspective from which least cost can be defined.

The Societal Perspective. From a societal perspective, global warming means that the true cost of using fossil fuels is much higher than apparent. To bring about the internalization of these costs, an energy or fossil carbon tax (or better, revenue-neutral incentive schemes) and/or regulatory action to limit fossil energy use would be justified. Such changes in price and/or regulatory requirements will make previously cost-effective energy-producing and energy-using equipment obsolete.

From a standard neoclassical perspective, cost-effectiveness is always measured at the margin. If it is cheaper at the margin to scrap the capital stock of energy-producing or energy-using equipment, this is the economically rational decision—even if these capital stocks were just built and purchased yesterday. Societal welfare maximization as formulated in neoclassical economics is not concerned with sunk costs.

The conflict over environmental policy in general, and over climate policy in particular, then, can be described as a conflict between the perspective of the owners of sunk investments and the societal perspective on least-cost efficiency. For example, it could be economically efficient policy for governments to phase in strict fuel economy standards even if these reduce the demand for gasoline at a rate significantly faster than the depreciation of existing refinery capacity. And a carbon tax on gasoline could be societally efficient even if this action were to reduce the resale value of existing automobiles.

Of course, global warming is not the only force that renders capital stocks obsolete. Technical innovation and international competition do the same. Moreover, even in the absence of global warming, there are large inefficiencies in existing energy-using capital stocks. A large portion of the demand for energy, and consequently of the energy supply business, is built on information and market barriers. These make it difficult for individual consumers to choose investments that minimize their energy service costs.[1]

As societies investigate options for reducing fossil fuel consumption, the general disequilibrium between the current prices of energy supplies and the cost and price of efficiency improvements is being more broadly recognized. Most importantly, the global warming issue is greatly strengthening the incentives for society to implement these "demand-side resources" through removal of market barriers and other policy action, because they afford opportunities to reduce greenhouse gas emissions *at negative societal cost*.

The Energy-Producer Perspective. Assume for the moment that the fossil fuel industry receives no subsidies to cushion it against a fossil phaseout. Least cost from the perspective of the fossil and utility industries means that the rate of fossil phaseout is slow enough not to create losses for their capital owners: existing investments in fossil supply and distribution would operate to the end of their anticipated life and still earn their expected returns. In effect, this producer's perspective would tie the rate of fossil fuel phaseout to the turnover rates of supply-side capital stocks.

If a more rapid restructuring were to be achieved, society could, of course, pay producers for their losses in the form of accelerated depreciation allowances. The fiscal losses incurred by such allowances could likely be recouped through revenues from fossil carbon taxes. Such subsidies would protect investors from capital losses and redistribute the burden to energy consumers at large.

Of course, even under a phaseout tied to supply-side capital stocks, capital owners in the fossil industries would have to bear an indirect cost, that is, the risks and uncertainties of having to diversify into other business activities. For instance, even a well-planned retrenchment could create impacts on the value of stocks in the financial markets.

---

[1]For a more detailed review of implicit consumer discount rates and underlying market barriers, see Krause and Eto (1988).

Government financial incentives could be required to make these risks acceptable to capital owners.

Fortunately, there would be a significant time window for diversification, since the lifetimes of existing capital stocks are measured in decades. Labor and regional governments, who can be considered part of the producer group in the present context, would have an extended period of time to make the necessary adjustments. Overall, the required rate of restructuring energy sector activities might be no greater than the rates of change forced on industries and society at large due to increased world economic competition. But unlike these larger pressures for structural change, the phaseout of fossil fuels could be made quite predictable by establishing clear agreed-upon national policy schedules for it.

The Energy-Consumer Perspective. Energy consumers, too, have made investments in fossil fuel using capital stocks that will be affected by government policies to internalize the risk of greenhouse warming. Policies such as fossil carbon taxes will make their equipment economically inefficient, or, if it is so already, even less cost-effective.

The business activities or living expenses of domestic and commercial and industrial energy consumers are not as centrally shaped by the price of fossil fuels as the operations of the energy suppliers. In fact, many of them are operating inefficient motors, furnaces, air conditioners, household appliances, and vehicles that would be worth scrapping: energy cost savings from new, more efficient devices would more than compensate for both the capital loss of early retirement and the extra capital cost of buying more efficient equipment.[2] In these instances, higher fossil fuel prices would help bring this disequilibrium to the attention of consumers, while technical regulations or targeted incentive payments would directly overcome the market barriers that keep consumers from obtaining energy services at least cost.

In some countries, these considerations have actually led to programs for the early retirement of existing capital stocks. In the United States, for example, a number of utility-sponsored efficiency incentive programs promote early retirement of inefficient existing end-use

---

[2]Any marginal demand-side or supply-side resource that can provide energy services at substantially lower cost than current systems could justify the early retirement of existing capital stocks. Often, these conditions are fulfilled in end uses where energy costs greatly outweigh the capital costs of end-use devices. Examples are commercial and industrial lighting and many electric motor applications. Here, even modest efficiency gains can make mass retrofits or wholesale replacements of existing equipment cost-effective.

equipment. In Europe, significant investments in the steel industry were made on a scrap-and-rebuild basis to increase energy efficiency and competitiveness.

Faced with a rise of fossil fuel prices due to government taxation, a much broader range of early retirement programs would be economically justified for the consumer. However, consumers, too, could require larger incentives for participating in such intervention than in programs in which the necessary adjustments are being made at the time of consumer replacement purchases.

The Continuity Rate. From a societal economic welfare-maximization perspective, the most desirable fossil phaseout is one that proceeds at the rate of logistic feasibility. But the more the rate of fossil phaseout exceeds the turnover rates of capital stocks, the greater will be the pressure faced by governments to provide compensatory incentives and redistributive fiscal programs. This pattern is clearly borne out by the experience of nations that underwent rapid structural change from international competition. Where such structural change occurred not only at the margin, that is, through a differentiation of growth rates among industries, but led in effect to the idling or premature retirement of existing productive capacity, governments have found it difficult to escape pressures for compensation from the taxpayers at large.

Even when fossil substitution is undertaken at no more than the rate of capital turnover, pressures to provide government compensation and resistance to phaseout policies are likely. And unlike the oil price hikes of the 1970s, global warming will create economic or psychological shock that could help overcome this resistance only with delay. For these reasons, a practical national target-setting guideline might be to phase out fossil fuels in approximate synchrony with the turnover rate of capital stocks. We call this rate the *continuity rate*.

This does not mean that governments should avoid faster action in individual industries and energy applications where logistic and political constraints so allow. On the contrary, wherever mass retrofits and scrap-and-rebuild programs would be justified on conventional economic grounds, these opportunities should be pursued as aggressively as possible, since they are likely to be counterbalanced by difficulties in other areas. The suggested continuity-rate guideline should be understood to represent the likely overall, average rate of fossil phaseout resulting from an aggressive but politically realistic program of climate stabilization. Within this umbrella rate of phaseout, individual policies and programs can trigger fossil fuel substitution at various speeds.

We therefore first investigate whether a rate of fossil fuel displacement closely tied to the turnover of capital stocks would be sufficient to meet climatic requirements. Should the continuity rate be incompatible with the global fossil carbon budget, we can then determine what faster action is required.

2. Phaseout Rates Based on Turnover Rates of Capital Stocks

The relevant capital stocks are those of the energy supply system on the one hand, and those of energy-consuming equipment and buildings on the other. One issue is whether the feasible rate of fossil fuel displacement from capital turnover on the supply side is significantly different from that on the demand side. Here, both the lifetimes of the relevant technologies and the degree to which the carbon intensity of new equipment can be increased during one replacement cycle are important. A further factor is the growth in energy service demand.

Turnover Rates. On the supply side, the most important long-lived capital stocks are power plants, with an economic life of about 30–40 years for a typical coal plant, and 20–30 years for other power plant types. An average life of 30 years is a good round number. This figure can also be used as a weighted average for supply-side facilities at large, including refineries, pipelines, and so forth.

On the demand side, vehicles, industrial processes, major appliances, and heating and air conditioning systems have technical-economic lives of 10–20 years. Building shells, which determine energy consumption, have much longer lifetimes: anywhere from 40 to more than 100 years. However, major renovations of buildings are due every 20–25 years, and significant improvements can be introduced at that time. A weighted average of about 20 years is a reasonable estimate.

Phaseout Functions. There is some confusion in the literature about how the percentage rate of fossil phaseout should be defined and calculated. A number of analyses connect target rates of fossil carbon emissions with the base-year emission rate through an exponential function. This choice of phaseout function is, of course, only one of any number of mathematical functions that could be used to connect the two emission levels.

An exponential function is, in many cases, inadequate for simulating turnover rates of capital stocks. It assigns very large absolute reductions to the early years of the phaseout, and progressively smaller absolute reductions to later years. This pattern only applies if capital stocks had been growing and then saturating in the years immediately preceeding the onset of the phaseout policy. If capital stocks were built up in an exponential growth path, the absolute number of retirements would be smallest in the early years, and would become exponentially larger in later years. If capital stocks were built up in a linear fashion, the absolute number of annual retirements would again grow in later years, though in a slower, linear fashion. If capital stocks had been at a steady-state level, a constant annual increment would be retired each year.[3]

In developing phaseout trajectories, we will be using a constant annual increment of fossil carbon reductions. In the initial years, we use smaller increments to reflect the logistics of starting up substitution programs. Compared to the exponential formula, a constant annual increment leads to higher levels of emission reduction in later years of the phase out cycle. This pattern is more plausible, since industries providing nonfossil technologies will be well-established in later years.

Growth of Energy Services. So long as some portion of marginal energy service growth is still being met by fossil fuels, this growth will counteract the phaseout from replacing existing fossil-fuel-based capital goods. To simplify our analysis, we assume that all marginal increases in energy services will be supplied by nonfossil energy sources, either directly, or through offsets provided by a limited number of mass retrofit programs.

For instance, if total housing stocks grow by 3 percent, we assume that the additional fossil fuels that would otherwise be required to supply these new homes will largely be supplied by a combination of improved efficiency and nonfossil energy supplies. To the extent that these new homes would still use some limited amount of fossil-based energy, the phaseout increment in existing stocks would have to be slightly larger. This could be achieved by focusing on retrofitting those buildings and measures that provide highly cost-effective energy savings even if the renovation cycle is somewhat accelerated.

Calculation of the Continuity Rate. If a crude model is used in which all facilities have the same lifetime rather than a distribution of life-

---

[3]Actually, each capital stock has a distribution of lifetimes. For a discussion of retirement functions in comparison to empirical data on appliances, see McMahon (1981).

times, and assuming steady-state conditions, about 3 percent of the world's supply investments would be replaced each year. Upon replacement, they could be switched to nonfossil energy sources, or to high-efficiency, low-carbon energy sources such as gas-fired cogeneration. As an approximate upper-limit range, we assume that such substitution could reduce fossil carbon releases by an absolute annual amount equal to about 2.6–2.9 percent of base-year releases.[4]

On the demand side, a weighted average lifetime of about 20 years would result in an annual rate of replacement of 5 percent per year. Since efficiency improvements eliminate only a portion of energy use, this 5 percent value needs to be corrected by the degree of efficiency improvement that can be introduced within a 20-year span. Assuming that product development, retooling, and commercialization limit the average improvement to 50 percent over 20 years,[5] efficiency improvements could contribute an annual reduction increment in carbon emissions of 5 percent/yr $\times$ 0.5 = 2.5 percent/yr of the base year value. Fuel switching at the point of end-use (e.g., from fossil-based electric heating to direct use of gas for heating) could add a further increment, particularly in those OECD countries where electric space conditioning is prevalent, and so could scrap-and-rebuild or mass retrofit activities. We round the demand-side continuity rate to a value of 3 percent to allow for such opportunities.[6]

Though only approximate, these figures show that roughly equal annual increments of fossil fuel displacement could be obtained from

[4]The basic alternatives are renewable energy sources, nuclear power, and cogeneration. Cogeneration technologies based on gas could contribute more than two-thirds and up to more than 90 percent of the carbon savings obtainable from a complete switch to nonfossil energy sources. If nonfossil and cogeneration/fuel-switching options contribute equally to capacity replacements, the annual percentage reduction in fossil carbon could be about (0.5 $\times$ 100% + 0.5 $\times$ 75%) $\times$ 3%/yr = 2.63% on a weighted average basis.
[5]This figure has emerged as a good rule of thumb from detailed studies of demand-side resource potentials. The longer-term potential for efficiency improvements in most end uses is 70–90 percent. While best technology would yield greater percentage savings in developing countries, institutional and implementation obstacles are also larger. We therefore use the same continuity rate for ICs and DCs. For further discussion, see for example, Krause et al. (1980), Chandler et al. (1988).
[6]Structural change is assumed to largely occur through lags in the positive growth rates of affected sectors, not rapid negative growth that would affect existing capital stocks. It is therefore subsumed in the treatment of energy service growth at large (see the subsection above).

supply-side investments and demand-side investments.[7] In a given year, then, fossil phaseout investments would displace an absolute amount of carbon equivalent to about 3 percent of the baseline releases. We further assume that this annual increment could continue to be displaced each year as long as carbon releases are still significant.[8]

3. Near-Term Constraints

Based on the above continuity rate, a 20 percent reduction in fossil carbon releases could be achieved in as little as $(20 \div 3)$ seven years. But in the near term, rates of change will of necessity be more limited. Any realistic goal for fossil fuel reductions must take into account the near-term momentum of past investment decisions. Also, supply facilities already under construction or in advanced planning stages will be largely fossil-based. And most of the world's buildings and equipment that are currently on order or being produced are not radically more efficient than existing stocks.

To this one must add initial transaction costs in reorienting energy policy. Here, factors other than purely economic concerns are important. Many industries that produce energy-using equipment will need to make significant product development and retooling efforts to manufacture more efficient equipment. In many cases, designers, engineers, and installers will need to be trained or retrained to implement the new technologies. Professional societies will need to embrace and certify more advanced techniques before they can be widely applied.

The energy supply companies will also need to be transformed. For example, utilities will need to change their mission from energy seller

---

[7]Note that the feasible rates of substitution from supply- and demand-side investments are not additive. Note also that the demand-side/supply-side equivalence only holds if nonfossil supply-side resources and efficiency-based demand-side resources are both equally available for market introduction and meet criteria of cost-effectiveness and nonclimatic environmental requirements equally well. Also, this equivalence only applies to the displacement rate while sufficient resource potentials are available. It does not describe the relative contributions of the two types of resources under least-cost criteria.

[8]To further illustrate the difference between the conventional growth rate approach and the annual increment approach used here, Tables 6.2 and 6.3 juxtapose both the annual increment and the exponential descriptors. Over time, as remaining fossil carbon releases drop to lower levels, the exponentially calculated negative growth rate of carbon releases becomes larger.

to energy service company or procurement broker of supply-side and demand-side resources. This, in turn, will require major regulatory reforms.

To implement a fossil fuel phaseout, new regulatory and administrative institutions will be needed. For instance, most any country has a ministry of energy, and many have a ministry of coal or nuclear energy, but none have ministries for energy efficiency or renewables. Governments will have to regroup to redirect the manner in which energy technology is developed and energy services are provided. Technology research communities will have to take on new missions.

Constraints in Developing Countries. If these near-term barriers seem formidable in the industrialized countries, they are at least as large in the developing countries. To a considerable extent, developing countries rely on imported or licensed technology. So long as industrialized countries have not widely commercialized efficiency improvements and renewables-based technologies that could be adapted to their special needs, the capacity and willingness of developing countries to switch to nonfossil energy systems are likely to remain limited.

In developing countries, fossil-fuel-saving technologies are often treated as luxury technology that is only applied to export items. A number of countries produce appliances and other items for export that are more energy-efficient than versions produced for the domestic market. In part, this reflects higher interest rates, which in turn are related to the debt problem and the associated capital flight from developing countries.

To remedy this situation, developing countries will need to implement incentives programs that overcome the barrier of higher first costs for energy- and capital-efficient technologies. They will also need to gain greater regulatory control over technology choices now made by corporations from industrial countries. The latter will need to move toward unifying environmental, efficiency, and cost-effectiveness standards through international agreements.

In some countries and in some areas of energy application, current Third World technological capacities could already support a fossil phaseout. Despite their flaws, initiatives such as the Brazilian fuel alcohol program show that the developing countries' technological dependence is by no means absolute. An important new technological option for developing countries is the building of dispersed efficiency/biomass/renewables-oriented local utilities to supply rural efficiency, electricity, and fuel needs.

4. Summary

The further we look into the future, the less important will be the inertia of these status quo constraints, and the more closely could carbon reductions proceed at approximately the continuity rate. The following propositions are hence made for defining continuity-based phase out trajectories:

- Once a fossil-phaseout program is in full swing, annual fossil carbon emissions from existing energy uses in both ICs and DCs could be reduced by an increment of approximately 3 percent of base-year emissions.
- So long as fossil carbon emissions from marginal energy service demand are less than 3 percent, economic growth can be achieved with declining fossil carbon emissions.
  - For example, a 4 percent growth in energy service demand could be met with a 1 percent reduction in total fossil carbon releases by reducing fossil carbon releases from existing stocks at the continuity rate of 3 percent and by supplying half of the marginal energy service demand, or 2 percent, with renewables and efficiency.
- In the near term, *developing countries* will need increasing amounts of fossil fuels. However, developing countries can contribute to climate stabilization by reducing fossil fuel consumption *in relative terms*, that is, by lowering the fossil carbon intensity of gdp (C/gdp ratio).
- Significant absolute reductions of fossil fuel consumption in developing countries could be achieved once efficient appliances, buildings, vehicles, and industrial plants have become widely available in the developing world, and once dispersed cogeneration and renewables-based supply technologies have reached commercial maturity and/or have been widely implemented through grass-roots efforts by local organizations. It may take 30–40 years until these conditions will be widely fulfilled.
- *Industrialized countries* will need several years to reverse current trends of rising fossil fuel consumption and set up the policies and programs that would allow absolute reductions.
- Beyond this start-up phase, lower growth in energy services, greater technology development capacities and commercialization infrastructures, and greater access to capital will make steep declines in IC fossil carbon emissions logistically feasible.

C. MILESTONES FOR THE GLOBAL FOSSIL PHASEOUT

1. What Regional Allocation of the Global Carbon Budget Should Be Used for Target Setting?

In analyzing near-term reduction milestones like the Toronto target, the allocational issue is overshadowed by logistic constraints in both industrialized and developing countries. Carbon emissions in the period until 2005 are going to be determined more by the inertia of current investments and policies than by the total carbon budget that will eventually be available to each region. For the larger reduction milestones, on the other hand, the regional allocation of the global budget becomes decisive.

In the following analysis, the global carbon budget is divided evenly among developing countries and industrialized countries. Given our discussion in Chapter 5, the 50–50 allocation is a simple compromise located approximately midway between the cumulative per capita formula, which would assign DCs 84 percent of the global fossil carbon budget, and the status quo carbon-release-based formula, which would assign them 28 percent.

However, at this stage of our analysis, a 50–50 regional split is only a working proposition. We next investigate whether such an allocation will produce a fair and manageable sharing of the task of fossil carbon reductions, given practical constraints on the rate of fossil fuel substitution in each group of countries.

2. Where Does the Toronto Target Fit within a Carbon Budget Approach?

In 1988, the Toronto World Conference on the Changing Atmosphere called upon governments to reduce world emissions of carbon dioxide from energy production by 20 percent by the year 2005. This call for a 20 percent reduction in carbon emissions by a high-level international conference is a historically significant and politically important event. It is the first acknowledgment by such a body of the need to *reduce* fossil carbon emissions. And it establishes a near-term target that affects current planning for major energy supply facilities and long-lived energy consuming goods.

Since the Toronto target is a political target, its relationship to the climate-stabilization goals as developed in this text needs to be clarified. It was pointed out in Chapter 1 that a single percentage milestone is conceptually insufficient for defining climatic targets. Additional, longer-term milestones should be specified for energy policy, since the effects of decisions made today on future energy-use patterns will be felt over much more than 20 years. Furthermore, while the target was formulated for the world as a whole, no specific guidelines were provided as to how the target might apply to developing and industrialized countries. Recent legislative initiatives in the United States and elsewhere have interpreted the target as directly applying to industrialized countries. By implication, developing countries would have to achieve a 20 percent reduction as well. Various nongovernmental organizations have called for larger reductions in the industrialized countries to accommodate growth needs in the developing countries.

To clarify these issues, we investigate the following questions:

- Does the Toronto target fall within the range of near-term milestones that would be required by, or be compatible with, the global warming analysis and carbon budget developed in this study?
- What reduction targets should or could apply to developing and industrialized countries?
- How feasible is the Toronto target given logistic constraints in each of the two regions?

The first question is easily answered by inspecting Figure 6.1: A 20 percent reduction by 2005 would be compatible with trajectories for realizing the 300-btC limit. By contrast, the Toronto target would not make sense under a 450-btC limit: in order to reach the steady-state release rate of 1 btC/yr late in the next century, the 450-btC emission trajectory would have to rise again after 2005 or the total budget could not be consumed. Such a trend reversal would be illogical and logistically incongruous.

3. The Toronto Target as a Global Milestone

We begin by applying the Toronto target as a global target. Using 1985 emissions as a baseline,[9] the Toronto target means a reduction of fossil carbon releases from 5.2 btC to about 4.2 btC in 2005. As a further limiting condition, we estimate the minimum fossil fuel needs of developing countries until 2005. We assume that developing countries will succeed in stabilizing their rising emissions by about the turn

---

[9]The Toronto target specified 1987 as the baseline year. The difference in emissions between the two years is small enough to be ignored in the present context.

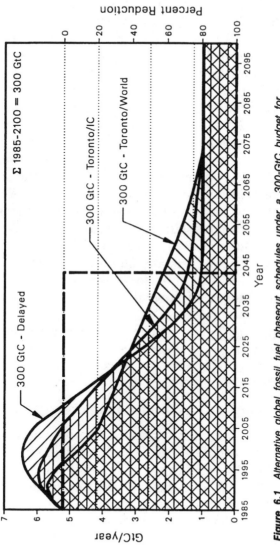

Figure 6.1 *Alternative global fossil fuel phaseout schedules under a 300-GtC budget for 1985–2100.*

of the century. In 2005, developing countries would release about 50 percent more fossil carbon than in 1985, or about 2 btC. The annual rates of fossil carbon emissions for this "DC 150-btC" trajectory (see Fig. 6.2) are shown in Tables 6.1 and 6.2.

This leaves 2.2 btC for the 2005 release from industrialized countries. To get there, industrialized countries would have to achieve a 43 percent reduction of their 1985 carbon emissions within a mere 20 years (not shown in Fig. 6.2 and 6.1). Of these, five years have already passed in which emissions increased further. This means a virtually instantaneous turnaround of present trends. On average, all growth in energy service needs and all facilities up for replacement between now and 2005 would have to be substituted by efficiency options and nonfossil supplies. Failure to do so in earlier years would have to be compensated for by crash programs in later years.

This is an extremely ambitious schedule. Could faster action in the developing world possibly provide relief? Even if the developing countries could somehow succeed in supplying *all* their marginal energy service needs from nonfossil resources starting in 1990, industrialized countries would still need to reduce their releases by close to 40 percent. In either case, the continuity rate would have to be exceeded by a substantial margin in the IC phaseout.

These quantifications indicate that

- Implementing the Toronto target as a global reduction milestone will require crash programs and accelerated depreciation of existing capital stocks.

4. The Toronto Target as a Milestone for Industrialized Countries

Assuming the same minimum carbon releases for developing countries, we might next examine a 20 percent reduction of industrialized country emissions. This application of the Toronto target to ICs only would result in 2005 *global* carbon releases that are no different from the 1985 level ("Toronto/IC" case—see Fig. 6.1 and Table 6.1). If IC programs are started immediately and succeed in holding emissions constant during 1990–1995, a period of ten years would be left until 2005 to implement a reduction of 20 percent. At an annual reduction increment of 2 percent of 1985 releases, this target would certainly be ambitious but would still fall within the bounds of the continuity rate (Table 6.3).

The conclusions are as follows:

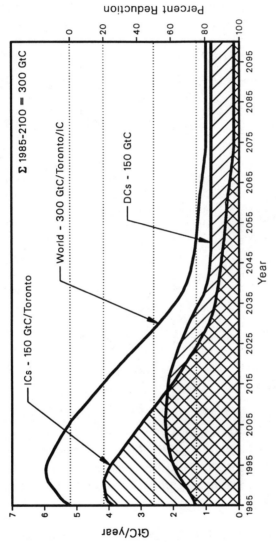

Figure 6.2 Application of Toronto target to industrialized countries.

TABLE 6.1 Fossil Carbon Releases, 1985–2100, by World Region under Various Phaseout Schedules: 300-btC Cumulative Budget and 450-btC Sensitivity Case

	1 ICs 150 btC Toronto/IC	2 ICs 150 btC delayed	3 ICs 225 btC sensitivity	4 DCs 150 btC Toronto/IC	5 DCs 225 btC sensitivity	1+4 World 300 btC Toronto/IC	2+4 World 300 btC delayed	3+5 World 450 btC sensitivity
1985	3.82	3.82	3.82	1.39	1.39	5.21	5.21	5.21
1990	4.15	4.20	4.20	1.65	1.65	5.80	5.86	5.85
1995	3.95	4.36	4.32	1.99	1.99	5.94	6.35	6.31
2000	3.48	4.27	4.08	2.16	2.24	5.64	6.43	6.32
2005	3.01	3.86	3.72	2.24	2.45	5.25	6.10	6.17
2010	2.52	3.08	3.30	2.21	2.62	4.73	5.29	5.92
2015	2.03	2.27	3.00	2.16	2.67	4.19	4.43	5.67
2020	1.63	1.57	2.74	2.00	2.63	3.63	3.57	5.37
2025	1.24	0.92	2.49	1.79	2.58	3.03	2.71	5.07
2030	0.91	0.43	2.24	1.53	2.53	2.44	1.96	4.77
2035	0.72	0.20	2.00	1.22	2.47	1.94	1.42	4.47
2040	0.62	0.15	1.80	0.97	2.37	1.59	1.12	4.17
2045	0.54	0.15	1.60	0.88	2.27	1.42	1.03	3.87
2050	0.47	0.15	1.40	0.85	2.17	1.32	1.00	3.57
2055	0.41	0.15	1.20	0.85	2.07	1.26	1.00	3.27
2060	0.36	0.15	1.02	0.85	1.95	1.21	1.00	2.97
2065	0.29	0.15	0.87	0.85	1.80	1.14	1.00	2.67
2070	0.21	0.15	0.72	0.85	1.65	1.06	1.00	2.37
2075	0.15	0.15	0.55	0.85	1.50	1.00	1.00	2.05
2080	0.16	0.15	0.42	0.85	1.28	1.01	1.00	1.70
2085	0.16	0.15	0.33	0.85	1.02	1.01	1.00	1.35
2090	0.16	0.15	0.25	0.85	0.75	1.01	1.00	1.00
2095	0.16	0.15	0.25	0.85	0.75	1.01	1.00	1.00
2100	0.16	0.15	0.25	0.85	0.75	1.01	1.00	1.00

- The Toronto target appears to be an appropriate and feasible near-term milestone for *industrialized* countries.
- With concomitant efforts in developing countries, an appropriate and feasible milestone for *global* fossil carbon emissions might be a return to about 1985 release rates by 2005.

In Table 6.1 and Figures 6.2–6.4, we show for the same Toronto/IC case how fossil carbon emissions from industrialized and developing countries would evolve in the years beyond 2005, given a 50–50 split of the 300-btC budget. The figures show that in the years beyond 2005, industrialized countries would have to continue to displace fossil fuels at rates close to the continuity limit (see Table 6.3).

TABLE 6.2 Required Fossil Carbon Reductions, gdp Growth, and Required Changes in C/gdp Ratios: Developing Countries

	Annual Change In C releases			Annual Change In C/gdp Ratios			
Historic changes							
1970–1986	1.00%			−2.03%			
1979–1986	−0.50%			−2.86%			
OECD 1979–86	−1.40%			−3.74%			
	Required change In C releases			Required change In C/gdp ratios (%/yr)			
				assuming a gdp growth of			
	Absolute btC	Index rel.to continuity limit (3% of 1985/yr)	Growth rate %/yr	1%	2%	3%	4%
150 btC Toronto/IC schedule							
1995–2000	−0.47	0.81	−2.50	−3.50	−4.50	−5.50	−6.50
2005–2010	−0.49	0.85	−3.50	−4.50	−5.50	−6.50	−7.50
2015–2020	−0.40	0.69	−4.30	−5.30	−6.30	−7.30	−8.30
2025–2030	−0.33	0.57	−6.00	−7.00	−8.00	−9.00	−10.00
150 btC delayed schedule							
1995–2000	−0.09	0.16	−0.40	−1.40	−2.40	−3.40	−4.40
2005–2010	−0.78	1.35	−4.40	−5.40	−6.40	−7.40	−8.40
2015–2020	−0.70	1.21	−7.10	−8.10	−9.10	−10.10	−11.10
2025–2030	−0.49	0.85	−14.10	−15.10	−16.10	−17.10	−18.10
225 btC sensitivity case							
1995–2000	−0.24	0.41	−1.10	−2.10	−3.10	−4.10	−5.10
2005–2010	−0.42	0.73	−2.40	−3.40	−4.40	−5.40	−6.40
2015–2020	−0.26	0.45	−2.80	−3.80	−4.80	−5.80	−6.80
2025–2030	−0.25	0.43	−3.10	−4.10	−5.10	−6.10	−7.10

The major milestones are summarized in Tables 6.4–6.6. The world as a whole would reach the −20 percent milestone by 2015, and the −50 percent milestone by 2030. Industrialized countries, on the other hand, would almost reach the −50 percent milestone in about 2015, or within about ten years of having achieved the 20 percent reduction goal. The −75 percent milestone would be reached within another 15 years, by about 2030.

These milestone years appear ambitious, but it should again be emphasized that they remain well within the bounds of the continuity rate (Table 6.3).[10] This leads to the following assessment:

- A 50–50 split of the global fossil carbon budget among developing and industrialized nations appears to yield workable and fair reduction milestones. It would push industrialized countries to fully

[10]Note that the increments in columns 2 of the table are averages for 5-year periods.

TABLE 6.3 Required Fossil Carbon Reductions, gdp Growth, and Required Changes in C/gdp Ratios: Industrialized Countries

Historic changes	Annual Change in C releases	Annual Change in C/gdp Ratios
1970–1986	5.60%	0.66%
1979–1986	3.70%	0.00%
Centrally Planned Asia 79/86	3.81%	-3.33%

	Required change in C releases			Required change in C/gdp ratios (%/yr)			
	Absolute btC	Index rel.to continuity limit (3% of 1985/yr)	Growth rate %/yr	assuming a gdp growth of			
				2%	3%	4%	5%
Toronto/IC schedule 150 btC[1)]							
1995–2000	0.17	—	1.65	-0.35	-1.35	-2.35	-3.35
2005–2010	-0.03	0.10	-0.30	-2.30	-3.30	-4.30	-5.30
2015–2020	-0.16	0.53	-1.50	-3.50	-4.50	-5.50	-6.50
2025–2030	-0.26	0.87	-3.10	-5.10	-6.10	-7.10	-8.10
225 btC sensitivity case							
1995–2000	0.25	—	2.40	0.40	-0.60	-1.60	-2.60
2005–2010	-0.04	0.13	-0.40	-2.40	-3.40	-4.40	-5.40
2015–2020	-0.15	0.50	-1.60	-3.60	-4.60	-5.60	-6.60
2025–2030	-0.18	0.60	-2.20	-4.20	-5.20	-6.20	-7.20

[1)]Toronto/IC schedule refers to reduction milestones for DCs under the 300-btC Toronto/IC schedule (see Table 6.1).

mobilize their technological, financial, and organizational capacities for phasing out fossil fuels without creating infeasible goals.

5. Sensitivity Analysis: Other Phaseout Trajectories for a 300-btC Budget

Abandoning the Toronto target entirely, we might ask: Could other viable 150-btC trajectories be defined for industrialized countries in which the onset of absolute fossil carbon reductions would be postponed into the next century? Unfortunately, such postponement would likely compound the logistical challenges. Since the "Toronto/IC" trajectory already proceeds close to the continuity limit in the near-to-medium term (Table 6.3), any scenario in which strong action is delayed until after the turn of the century is likely to require accelerated replacement of capital stocks or scrap-and-rebuild programs beyond those that would be economically advantageous anyway.

For example, suppose the Toronto/IC target were replaced with the more lenient goal of a return to 1985 levels by 2005, with a shallow peak in between. This variant, which is not plotted in the figures, is labeled "150 btC delayed" in Tables 6.1 and 6.2. Such a delay in-

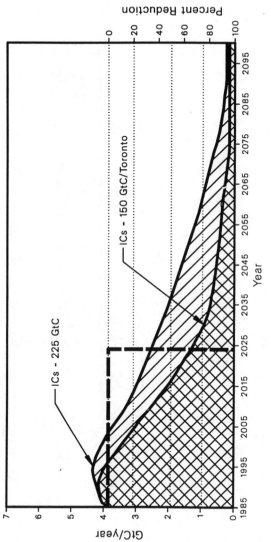

Figure 6.3 *Fossil fuel phaseout schedule for industrialized countries.*

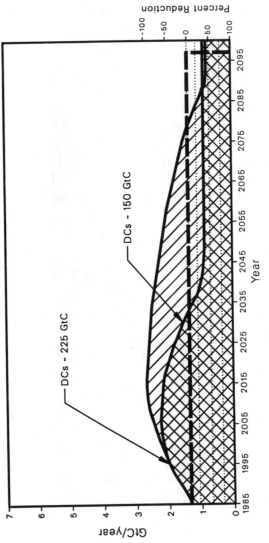

Figure 6.4 Fossil fuel phaseout schedule for developing countries.

TABLE 6.4 Reduction Schedules for Global Fossil Carbon Releases: 300-btC Limit and 450-btC Sensitivity Case[a]

Maximum release rate, btC/yr		6	6.5	6.3		
Index (1985 =1.00)		*1.15*	*1.25*	*1.21*		
% of C budget burned by 2005		39%	43%	27%		
% of C budget burned by 2030		75%	71%	57%		

Milestone	btC/year	1 300 btC Toronto/ IC Year	2 delayed Year	3 450 btC Sensitivity Year	4 difference (2–1) years	5 difference (3–1) years
Year of peak release rate		1994	1999	2000	5	6
return to 1985 level (0%)	5.21	2005	2010	2023	5	18
−20%	4.16	2015	2017	2040	2	25
−50%	2.60	2029	2026	2066	−3	37
−75%	1.30	2050	2038	2086	−12	36

[a]Historic release rates based on Marland et al. (1988), excluding cement.

creases by about 25 percent (ten percentage points) the share of the 150-btC budget that is used up before 2005. To make up for this, reductions between 2005 and 2030 would have to be steeper: a drop of 90 percent would have to be achieved in these 25 years, compared to 70 percent in a comparable period for the Toronto/IC trajectory. As shown in Table 6.3, the continuity limit would have to be exceeded.

TABLE 6.5 Reduction Schedules for Fossil Carbon Releases from Industrialized Countries: 150-btC Limit and 225-btC Sensitivity Case[a]

Maximum release rate, btC/yr		4.15	4.36	4.36		
Index (1985 =1.00)		1.09	1.14	1.14		
%of C budget burned by 2005		53%	60%	37%		
%of C budget burned by 2030		84%	92%	69%		

Milestone	btC/year	1 150 btC Toronto/ IC Year	2 delayed Year	3 225 btC Sensitivity Year	4 difference (2–1) years	5 difference (3–1) years
Year of peak release rate		1991	1995	1995	4	4
return to 1985 level (0%)	3.82	1996	2005	2003	9	7
−20%	3.06	2005	2010	2013	5	8
−50%	1.91	2016	2017	2037	1	21
−75%	0.96	2030	2040	2062	−10	12

[a]Historic release rates based on Marland et al. (1988), excluding cement.

TABLE 6.6 Fossil Carbon Reduction Schedules for Developing Countries: 150-btC Budget and 225-btC Sensitivity Case*

Maximum release rate, btC/yr			2.25	2.67	
Index (1985 =1.00)		*1.62*	*1.92*		
% of C budget burned by 2005			26%	10%	
% of C budget burned by 2030			59%	18%	
Milestone	btC/year		150 btC Toronto/ IC Year	225 btC Sensitivity case Year	difference (2–1) years
Year of peak release rate			2026	2014	*–12*
return to 1985 level (0%)	1.39		2005	2079	*74*
–20%	1.11		2015	2084	*69*
–50%	0.70		—	—	—
–75%	0.35		—	—	—

*Historic release rates based on Marland et al. (1988), excluding cement.

6. Summary

Each of the global 300-btC trajectories contains subperiods in which the rate of change of carbon releases would appear particularly ambitious from the viewpoint of past developments. In the Toronto/IC trajectory, it is the rapid turnaround of present global trends by the mid-1990s that presents the greatest challenge. Subsequent rates of reduction appear less formidable.

In the "delayed" phaseout scenario, the timing in the initial 15–25 years may reflect more closely the difficulties of concerted international action. But inertia and procrastination have their price. The rate of reduction required in subsequent decades—an 80 percent reduction within about 35 years—appears very rapid, and scrap-and-rebuild programs would likely be required.

Note that the areas of concern in each of these extreme cases are somewhat different in nature. For the early phaseout, it is principally the world community's ability to come to consensus and act quickly and decisively that is in question. For the delayed phaseout, it is the rapid implementation of large-scale energy substitution processes in later years that requires greatest attention.

Despite considerable differences between the two 300-btC scenarios in the initial years and decades, the decision-making latitude between the two scenarios is not large: The zero and −20 percent milestones are only seven to ten years apart (see Table 6.4). Of course, other trajectories could be envisioned, but whatever leeway could emerge within the continuity rate is not much more than about ten years.

D. ARE THERE ANY OTHER OPTIONS?

1. Could an As-Yet Undiscovered Energy Technology Provide Relief?

The enormous changes brought about by modern physics, biology, computer science, and electronics continue to fuel hopes that the discovery of a new energy technology or energy source could at once do away with current economic and environmental constraints on energy use. Fundamentally new energy sources such as cold fusion or solar-based catalytic photolysis of water may well materialize, but would they help solve the predicament of climate warming?

Here, it must be remembered that any new energy source will be initially constrained by prototype development, commercialization, and market acceptance hurdles. For any energy source or technology not in existence today, this means a delay of at least 20 years before commercial market penetration can begin, and probably much longer. And unless such new sources are very inexpensive, their market penetration will be further limited by the same process of turnover of capital stocks as the penetration of currently available technology options.

In view of this, any policy that ignores the climatic limits and milestones developed in this chapter really banks on the discovery of universally applicable new energy sources *by the turn of the century*. It also implies that such energy panaceas would have low costs compared to present energy costs and prices.

Only if one is sure that both these conditions will be met could a policy course be justified that deviates from the above phaseout schedules. If such an energy technology were, for example, discovered in 2010 and commercialized in 2030, but fossil fuel consumption were to continue more or less unabated between now and then, the climatic damage would according to present scientific understanding already have been done.

For these reasons, the backbone of climate-stabilizing energy strategies must be the implementation of currently available low-carbon or nonfossil technology options. Accelerated technology development is surely needed to diversify and facilitate the phaseout process, but it cannot take the place of moving aggressively to implement currently available options. There are ample fossil-carbon-displacing technologies, either commercial or near-commercial, to realize fossil carbon reductions at the pace outlined above. In fact, the supply of carbon reduction options appears to be sufficient for policymakers to be selective in their deployment and to rely on those resources that can provide energy services at least economic, environmental, and social cost.

The conclusion is that

- Climate-stabilizing energy policies should include intensive development of new supply and demand-side technologies to diversify technology options and reduce costs. However, results from such research and development efforts cannot substantially alter the near-term imperatives for action based on existing technologies.

The inertia of existing structures in the energy supply system, in energy-consuming capital stocks, and in the political economy that maintains these structures cannot be overcome with technological measures alone.

2. Sensitivity Analysis: 450-btC Limit

From the standpoint of implementation, the outer climatic limit of 450 btC is easier to achieve than the 300-btC base. Unfortunately, this carbon budget is much less attractive from a climatic point of view. As pointed out before, it is based on best-case estimates about the global carbon cycle. These estimates appear even less suitable as a policy basis in view of the most recent atmospheric carbon dioxide concentration measurements: in 1988, the airborne fraction of fossil carbon emissions increased to 90 percent, compared to the 55 percent average of measurements over the last three decades. This increase could be due to a loss of CO_2 absorption capacity in the world's oceans (MacDonald 1989).

With this perspective in mind, we construct a phaseout schedule as shown in Figure 6.5 and Tables 6.1, 6.4, and 6.5. (column 3). To hedge against climatic risks, we assume broadly similar near-term changes in carbon release patterns as in the 300-btC "Toronto/IC" case. What relief in the reduction schedules is provided by relying on the 450-btC budget is largely allocated to later decades. In the next three to four decades, then, the main focus in this scenario would be not so much to achieve rapid absolute reductions, but to supply all energy service growth from efficiency and other nonfossil sources. Serious cuts in absolute emissions would begin in the second quarter of the next century.

Table 6.1 shows that global emissions peak around the turn of the century at about 21 percent above the 1985 baseline. Carbon releases

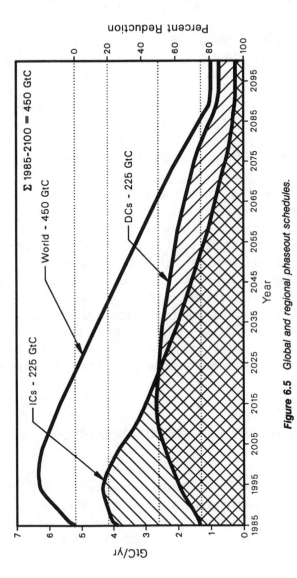

Figure 6.5 Global and regional phaseout schedules.

then decline more or less linearly at a modest pace of about 1 percent per year. World carbon releases would return to 1985 levels between 2020 and 2025. The −20 percent milestone would be reached in 2040, the −50 percent milestone in about 2065, and the −75 percent milestone around 2085.

Under a 50–50 split, fossil carbon releases in the developing countries would grow until around 2015, peaking at twice the 1985 level. The fossil carbon growth rate would be 2.8 percent over the next 25 years, or about half the 1970–1986 average (Table 6.2). Assuming a growth rate of 3 percent in per capita gdp, and population growth rates of 1.6 percent, carbon emissions in these early years would grow about $2.8/(3 + 1.6) = 60$ percent as fast as the gdp.[11] The "DC 225-btC" scenario would thus still represent a substantial fossil phaseout effort compared to the roughly lockstep pattern of the 1970–1986 period.

This lower DC fossil carbon growth between 1990 and 2015 articulates not just climate considerations, but developmental imperatives as well. Lowered C/gdp ratios would be economically rational for developing countries in view of their need to be competitive in international markets, as already discussed in Section D of Chapter 5.

Furthermore, many carbon-saving efficiency improvements and renewables investments are least-cost solutions on the basis of capital scarcity, foreign debt, and foreign-exchange earnings considerations alone. The climate-stabilization requirements incident on developing countries are therefore not necessarily an additional burden, but would blend well with other requirements derived from overall developmental goals.

E. C/GDP RATIOS: HISTORIC VERSUS REQUIRED FUTURE CHANGES

In Sections B–D, we relied on the turnover rates of capital stocks as a guideline for scenario development. Because most economic and energy planners employ the gross domestic product as an indicator, we also discuss the phaseout scenarios in terms of changes in the fossil carbon/gdp ratio.

---

[11]The 3 percent per capita figure represents roughly the mean between the growth rates of the 1965–1975 and 1975–1985 periods. The population growth rate reflects the U.N. medium projections discussed in Chapter 5.

1. Historic Data

Despite the continued general rise in carbon emissions over the last 20 years, the world economy has actually become more fossil carbon efficient. Figure 6.6 shows the global trends in energy/gdp ratios and C/gdp ratios since 1970.[12] Over the 16-year period until 1986, there has been a significant and steady decrease in both ratios. Due to investments in hydro and nuclear power, the carbon intensity of world economic product decline somewhat faster than the energy intensity of world gdp.

The figure shows that while gdp grew by more than 3 percent per year, carbon emissions trailed this rate considerably, starting in about 1973. Beginning with the first OPEC price hike, carbon releases lagged more and more behind economic growth. In the period from 1979 to 1985, world carbon emissions remained constant (see Chapter 5), while gdp still grew by roughly 3 percent per year. As a result, the carbon intensity of gdp dropped by about 20 percent between 1970 and 1986. Most of this drop was realized after the second oil price shock in 1979. Since 1985, carbon releases have been on the rise again, and improvements in C/gdp ratios appear to be small.[13]

Annual growth rates in C/gdp ratios for the world as a whole were about −2.0 percent for the 1979–1986 period, and −1.4 percent for the 1970–1986 period, compared to a growth in world gdp of about 3.0 percent and 3.6 percent per year, respectively.

Figure 6.7 shows that there were large differences among the various world regions. Clearly, the greatest changes in C/gdp ratios occurred in the OECD (Organization for Economic Cooperation and Development) countries (developed market economies). Since these nations constituted about 53 percent of world energy use in 1986, OECD patterns strongly shape the figures for the industrialized countries as a whole, and also dominate the global trends. The carbon intensity of

[12]Carbon release data from Marland et al. (1988). World gdp data were kindly provided by Kent Anderson (1989) and are based on World Bank data (1988) for market economies and on data from IISS (1987) for centrally planned economies. These data should be considered approximate only. Estimates of gdp growth are often distorted by exchange rates, accounting differences, and other data inconsistencies, including the presence of sizable underground economies in many nations. Uncertainties are probably greatest in estimates for low-income countries and in comparing the growth figures of centrally planned economies with those of OECD countries.

[13]Note that changes in C/gdp ratios stem from efficiency investments, structural changes in the economy, and behavioral factors. The relative contributions of these components vary from country to country.

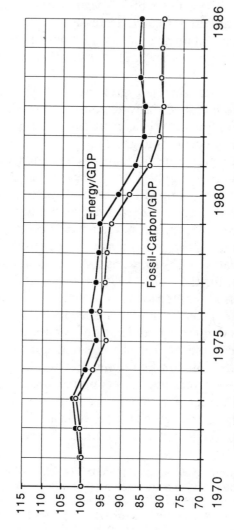

Figure 6.6 World energy/gdp and fossil carbon/gdp ratios, 1970–1986 (1970 = 100).

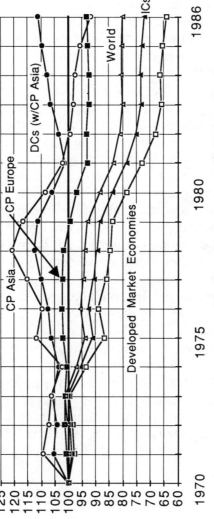

Figure 6.7 Evolution of carbon/gdp ratios by world region, 1970–1986 (1970 = 100).

gdp in the developing world increased slightly over this period. Likewise, carbon intensity in the centrally planned economies ended the period little changed from the 1970 value.

An indication of the potential for achieving economic growth in the Third World at lowered carbon intensity is provided by China (Chandler 1988). As shown in Figure 6.7, the C/gdp ratio in that country dropped by 23 percent between 1978 and 1986, a period of rapid growth and modernization efforts. C/gdp ratios changed at an average rate of −3.3 percent over the 1978–1986 period. This performance was significantly better than the world average, and fell between that of the ICs as a whole and that of the OECD (see Tables 6.2 and 6.3). About 20–40 percent of this remarkable drop appears to have been the result of structural changes toward consumer-goods industries. But most of the change seems to be attributable to specific policy measures taken since 1978 to promote more energy-efficient capital stocks and active energy-management measures to increase energy productivity and efficiency in the economy (Liu 1989).

One might suspect that this high rate of improvement was feasible only because Chinese capital stocks are even more energy-inefficient than those of other developing countries. This may be true in comparison to some of the higher-income developing countries. But a recent comparison of Chinese energy intensities *on a physical unit basis* with those in another major low-income country, that is, India, revealed that the two nations had very similar technological efficiencies (Lu et al. 1987). Differences in terms of the energy/gdp ratio exist principally because China's industrial mix is dominated more by heavy industry than that of India. Also, international statistics on China's gdp may be distorted by the prevalence of administrative prices in that country.

Also, China is not the only DC that managed to break its historic lockstep pattern and lower C/gdp ratios. Brazil's C/gdp ratio also dropped by about 20 percent between 1970 and 1986. Again, deliberate policies such as the fuel alcohol program contributed to this outcome.[14]

---

[14]Based on gdp figures of the World Bank (1988) and of the International Energy Studies Group at Lawrence Berkeley Laboratory. Both data series indicate that the C/gdp ratio did not really drop until after 1979. Thereafter, the annual percentage reduction of the C/gdp ratio closely matches that achieved in China. However, this pattern could also be due to discontinuities in gdp accounting and normalization methods (Sathaye 1989).

2. Future Requirements

In Tables 6.2 and 6.3 we show how C/gdp ratios would have to evolve under the various phaseout scenarios. A matrix of plausible values for future gdp growth rates is provided. As a point of reference, the tables also show the historic C/gdp growth rates. Note that industrialized countries achieved a gdp growth rate of about 3.3 percent per year in the decade from 1975 to 1985, while the rate was close to 5 percent for the developing countries.

Annual percentage rates of change in the C/gdp ratio are most informative for the near term, where they can be compared with recent experience. Over the longer term, significant changes occur in the C/gdp base value that make comparisons with historic growth rates less meaningful. Here, the continuity limit index in Tables 6.2 and 6.3 should be used for orientation.

Based on the 1979–1986 data and the growth matrices in the two tables, the following observations for the near term (the next 10–20 years) can be made:

- If *industrialized countries* continue to grow at 2–3 percent per year and are to meet the Toronto/IC target, they will need to increase, by about 50–120 percent and within the next few years, the rate of C/gdp reduction achieved in the 1979–1986 period.
- Even under the more lenient 225-btC schedule, industrialized countries will need to exceed the 1979–1986 carbon substitution rate (by as much as 50–80 percent) in the late 1990s (see the matrix in Tables 6.2 and 6.3).
- If developing countries grow at a similar per capita rate as in the past, they will need to achieve a moderate decoupling of carbon releases and gdp growth beginning in the 1990s.[15] In the first decade of the next century, they will need to match or exceed the rate of reduction in C/gdp ratios of about 3.3 percent per year achieved in the People's Republic of China during 1979–1986.

---

[15]Assuming continued per capita rates of about 2.5 percent/yr plus projected population growth of about 1.5 percent/yr (see Chapter 5), the gdp growth rate would be 4 percent/yr. According to the matrix of Table 6.2, C/gdp growth rates in the late 1990s would have to be −2.35 percent/yr under the Toronto/IC target, compared to 0 percent/yr for the DCs in 1979–1986.

TABLE 6.7   Allocation of the IC Fossil Carbon Budget on the Basis of 1986 Per Capita Fossil Carbon Releases

	1986 Release rate tons/cap-yr [1]	Percent of IC average	1985-2100 Fossil C allowance [2] btC	Years left at 1986 release rates
Portugal	0.79	23	1.26	155
Spain	1.28	38	4.78	96
Yugoslavia	1.49	44	2.88	83
Greece	1.62	47	1.23	77
New Zealand	1.63	48	0.41	75
Italy	1.65	49	7.06	74
Israel	1.68	49	0.53	73
Switzerland	1.79	53	0.80	70
France	1.79	53	6.84	70
Austria	1.93	57	0.94	65
Sweden	1.95	57	1.04	64
Iceland	1.97	58	0.03	62
Hungary	1.98	58	1.31	62
Japan	2.11	62	14.99	59
Norway	2.14	63	0.52	58
Ireland	2.15	63	0.44	57
Romania	2.41	71	2.83	51
Netherlands	2.41	71	1.80	51
Belgium	2.68	79	1.22	46
South Africa	2.78	82	3.99	43
United Kingdom	2.94	86	7.00	42
Finland	3.02	89	0.60	41
Federal Republic of Germany	3.07	90	7.51	40
Poland	3.32	97	4.63	37
Denmark	3.34	98	0.63	37
USSR	3.59	105	34.69	34
Bulgaria	3.60	106	1.11	34
Australia	3.85	113	1.97	32
Canada	4.09	120	3.16	30
Czechoslovakia	4.21	124	1.91	29
United States of America	5.01	147	29.81	25
German Democratic Republic	5.50	161	2.05	22
Luxembourg	6.42	188	0.05	19
TOTAL IC AVERAGE	**3.22**	**100**	**150**	**47**

[1] 1986 release rates from Marland et al. (1988), including cement.
[2] Allocation shares calculated on the basis of 1986 U.N. population data.

3. Carbon Budgets and Phaseout Requirements for Individual Industrialized Countries

So far, we have not differentiated among industrialized economies, though these differ greatly in their fossil carbon intensity. Table 6.7 shows a comparison of per capita 1986 carbon releases, and cumulative carbon budgets by country based on 1986 population shares. The use of current population shares seems practical, since the population growth rates among industrialized countries are small and by and large convergent.

At 1986 release rates, these cumulative fossil carbon budgets are equivalent to anywhere from about 100 years of fossil fuel consumption to as little as about 20 years. Countries at the lower end of this range would have to phase out fossil fuels at a rate significantly faster than the continuity rate. Alternatively, they would have to buy or negotiate larger emission budgets for themselves than computed from their population share.

The figures in Table 6.7 indicate that a significant amount of emission-rights trading could occur among the ICs. Among countries with relatively larger emission budgets are Spain, Italy, Switzerland, France, Sweden, and Japan. The United Kingdom, the Netherlands, Romania, and the FRG occupy the middle ground. Besides the United States, other countries with small carbon budgets would be the German Democratic Republic, Czechoslovakia, and Canada.

F. SUMMARY AND CONCLUSIONS

The following overall conclusions can be made:

- A global fossil carbon budget of 300 btC appears to be sufficient to allow an orderly phaseout of fossil fuels.
- In broad terms, the critical period in which the phaseout of fossil fuels will need to be largely accomplished is between now and the middle of the next century.
- The 300-btC limit could be realized within the limits of the continuity rate of about 3 percent per year, if decisive fossil-substituting action is set in motion within the decade of the 1990s.
- Intensive development of technologies is needed, but overreliance on major technological breakthroughs should be avoided, due to the need to achieve large fossil carbon reductions in the near and medium term.

The following milestones should guide the global fossil phaseout (Table 6.4):

- By 2005, global carbon releases should have returned to the 1985 level.

- The 20 percent global reduction milestone should be achieved by about 2015, or within about 25 years (base year 1990).
- A 50 percent global reduction should be achieved by about 2030, or within 40 years.
- A 75 percent reduction should be reached by about 2050 or within 60 years.

Regional phaseout goals might be formulated as follows:

- Reduction milestones based on allocating 150 btC each to industrialized and developing countries is probably close to an optimal compromise between international equity and practical feasibility.

Based on this allocation, the following targets should be adopted (Tables 6.5 and 6.6):

- The Toronto target should be understood as a target for industrialized countries.
- ICs should achieve a 20 percent reduction by 2005, a 50 percent reduction by about 2015, and a 75 percent reduction by about 2030.
- Developing countries as a group should aim to stabilize their fossil fuel consumption early in the next century, and limit increases in the meantime to about 50–100 percent above 1985 levels.
- The long-term level of fossil carbon releases from DCs should be about half their current release rate. This level should be reached by about 2050.

Within each group of countries, the application of per capita equity formulas would lead to further differentiation of phaseout milestones. Part in terms of the fossil carbon intensity of gross domestic product, these targets would read about as follows:

- In industrialized countries, near-to-medium-term reductions in C/gdp ratios will need to exceed 1979–1986 rates by as much as a factor of 2.
- Developing countries as a whole will need to overcome their lock-step C/gdp pattern and achieve rates of carbon substitution of about 3 percent per year, which are similar to those realized in developing countries with the most proactive energy policies during 1979–1986.

Having established these approximate targets, the key question becomes: Which mix of substitution technologies and energy policies holds the greatest promise for achieving fossil carbon reductions in a timely manner, and with the least economic and other costs? An exploration of this question is reserved for another publication (Krause et al. 1991). The next and final chapter explores the implications of our analysis for the shaping of an international climate convention.

REFERENCES

Anderson, K. (1989), "Estimate of World gdp Data Based on Economic Growth and Exchange Rate Data," Institute for Policy Studies, University of California, Berkeley, unpublished manuscript.

Chandler, W. U. (1988), "Assessing Carbon Emission Control Strategies: The Case of China," *Climatic Change*, Vol. 13, No. 3, pp. 241–265.

Chandler, W. U., H. S. Geller, and M. R. Ledbetter (1988), *Energy Efficiency: A New Agenda*, American Council for an Energy-Efficient Economy (ACEEE), Washington, D.C.

IISS (1987), *The Military Balance*, International Institute for Strategic Studies, London.

Krause, F., H. Bossel, and K. F. Müller-Reissmann (1980), *Energiewende*, S. Fischer Verlag, Frankfort.

Krause, F., and J. H. Eto (1988), *Least-Cost Utility Planning: A Handbook for Public Utility Commissioners, Vol. 2: The Demand Side: Conceptual and Methodological Issues*, National Association of Regulatory Utility Commissions (NARUC), Washington, D.C.

Krause, F., et al. (1991), *The Cost of Carbon Reductions: A Case Study of Western Europe*, IPSEP, El Cerrito, CA, November.

Liu, F. (1989), "Energy Efficiency and Economic Development in China," Lawrence Berkeley Laboratory, Applied Science Division, Energy Analysis Program, Berkeley, CA, July, unpublished manuscript.

Lu, Y. Z., et al. (1987), *The Study of China's Energy Utilization and Policies for Comparison with India*, Annual report of 1987, Institute of Techno-Economics and Energy Systems Analysis, Tsinghua University, Peking.

MacDonald, G. (1989), Presentation at the IEA Expert Seminar on Technologies for Controlling Greenhouse Gases, OECD, Paris, Apr. 12–14, MITRE Corporation, McLean, VA.

Marland, G., et al. (1988). *Estimates of CO_2 Emissions from Fossil Fuel Burning and Cement Manufacturing using the United Nations Energy Statistics and the U.S. Bureau of Mines Cement Manufacturing Data*, Carbon Dioxide Information Analysis Center (CDIAC) Oak Ridge National Laboratory, Oak Ridge, TN.

McMahon, J. E. (1981), *Residential End-Use Demand Modeling: Improvements to the ORNL Model*, LBL-12860, Lawrence Berkeley Laboratory, Berkeley, CA.

Sathaye, J. (1989), Personal communication, Lawrence Berkeley Laboratory, Berkeley, CA.

World Bank (1988), *World Development Report*, The World Bank, Washington, D.C.

7

A Global Compact on Climate Stabilization and Sustainable Development

A. CLIMATE STABILIZATION AND SUSTAINABLE DEVELOPMENT

1. The Need for a Broad North–South Compact

Much of the current climate warming debate still proceeds along the narrow lines of conventional air pollution abatement policy. But climate stabilization is an entirely different challenge. The greenhouse effect is driven by a confluence of environmental impacts that have their source not only in the nature of human resource use, but also in the nature of the current international economic order. Climate stabilization therefore requires a comprehensive turn toward environmentally sound and socially equitable development—in short, an unprecedented north–south compact on sustainable development.

Like no other environmental crisis, the threat to the world's climate drives home the point that all nations truly are in one boat—spaceship Earth. Industrialized countries, who long considered themselves the lucky winners in the global race for turning natural resources into economic wealth, cannot safeguard their economic future over the long term unless the driving forces of global climate change and environmental destruction in the developing world are brought to a halt. Most Third World nations, on the other hand, are caught in a short-term struggle for economic and physical survival. This struggle tragically pitches them against the long-term maintenance of the very environment on which their—and the world's—future depends. So long as this situation persists, developing countries cannot turn their attention to issues of sustainable development and climate stabilization.

In this situation, simple appeals to developing countries to join the industrialized countries in reducing their greenhouse gas emissions are not only naive but morally indefensible. No approach to climate stabilization can be successful unless it simultaneously rekindles economic growth and social progress in the Third World. An international effort to solve the Third World development debacle has to be part and parcel of overcoming the global environmental crisis.

2. A Rough Outline

The basic propositions of the needed global compact might be stated as follows:

1. Developing countries and industrialized countries will agree to undertake a joint venture to protect the global environment (in particular, the global climate) and to subordinate all national policies to that goal.
2. DCs and ICs will agree to unprecedented emission limits and other internationally defined environmental constraints on their national development paths.
3. The ICs will engage in an unprecedented environmental initiative to contain and reduce their disproportionate consumption of global environmental resources. They will treat the debt and basic needs crisis in the developing world, and the related environmental destruction, as an externality of their own development to date. Specifically,
   - ICs will drastically cut their greenhouse gas emissions in the quickest way possible.
   - ICs will agree to provide the necessary debt and international trade relief to restart Third World economic growth.
   - ICs will approach future technology development and technology transfer and trade policies from a global environment perspective.
4. The DCs will accept emission constraints, which ICs did not face during their development, and find alternative development paths that will not repeat and magnify the environmental impacts of past IC development. Specifically,
   - DCs will intensify their campaigns for population control.

- DCs will mobilize their communities for sustainable economic growth through land reforms and other political reforms and social policies that support local initiative and self-reliance.
- DCs will replace environmentally unsound price subsidies, notably for energy and other commodities, with efficiency subsidies.
- DCs will be selective in the adoption of technologies from ICs and will emphasize technical and infrastructural solutions that are better adapted to their social and economic needs and to their renewable base.

5. This acceptance of climate stabilization responsibilities by the DCs is tied to the following conditions:
   - DCS will be given prompt debt and trade relief to restart their economic growth.
   - ICs will provide aggressive leadership by example, notably in the area of industrial and fossil emission reductions.
   - The ICs will provide DCs with capital and technology transfers to allow them to leapfrog into growth based on highly efficient, nonpolluting technologies.
   - DCs will be given support to strengthen their own technology development capabilities and to manufacture high-efficiency, nonpolluting technologies themselves.
   - Beyond agreed-upon environmental and other international targets, the sovereignty of the DCs to pursue their own development paths will be respected.
   - DCs will have substantial representation and say in international funding decisions.

6. Effective use of international capital flows could be ensured through the following arrangements:
   - Funding of internationally sponsored projects and policies will be tied to specified ecodevelopmental conditionalities (see below).
   - Beyond a basic level of debt reduction and capital and technology assistance, additional incentives will be offered to nations exceeding certain energy and other resource efficiency indices, as well as environmental and population control targets.

7. The necessary capital for this global scheme could be obtained as follows:
   - The major industrialized nations will gradually reduce their military spending.

- Worldwide efficiency improvements in energy use will free capital for other purposes.
- Agricultural systems based on sophisticated low-input techniques could improve capital productivity worldwide, and both capital and labor productivity in Third World agricultural systems.
- Expanded global markets for high-efficiency technologies and renewables-based energy systems will provide a major stimulus to the world economy.

The above outline appeals to common sense, and, at the same time, it is thoroughly utopian. The profoundness of change in human consciousness and culture required to implement such a compact cannot be underestimated.

In proposing the above compact, we do not suggest that the world could somehow buy itself out of its present predicament merely by redeploying established means within the prevailing model of consumption-oriented industrialization and international trade. The nations of the world will have to come to terms with the structural impediments to economic development, both within the current international order and within the political and socioeconomic structures of each country. And changes in the development model pursued by Third World nations will not only require environmentally sound modern sector technologies but also environmentally sound intermediate technologies that are matched to the employment needs and capital sources of developing countries.

A common characteristic of countries that successfully industrialized in the past was their ability to generate an ever-greater forward and backward interlinking between the agricultural sector, the capital-goods sector, and the intermediate-goods and consumer-goods sectors. In this process of intensifying interlinkages, a significant part of industrial investments was aimed at simple products satisfying the basic needs of the large masses of lower-income consumers. This production, in turn, often drew in significant part on dispersed informal sector industries using intermediate technologies to provide semifinished inputs.[1]

This is the opposite of current patterns in most developing countries, many of which are still caught in colonial patterns of dependence

---

[1] One of the most cogent analyses of the commonalities of the successful industrializing nations, who showed large differences in resource endowments and trade strategies, is found in Senghaas (1982).

on exports of one or a few agricultural commodities. Others first attempted import substitution industrialization aimed at luxury-goods production and then tried export-oriented industrialization. Economies based on both models accrued debts far faster than internal interlinkages or distributed domestic employment and income effects took hold. As a result, most developing countries are experiencing a worsening—and environmentally highly destructive—dualism between isolated modern industrial enclaves and vast impoverished regions.[2]

In the economies of successful industrializers, on the other hand, the progressive interlinking of the various domestic production sectors created resilience and broad-based dynamism. It was usually accompanied by a reasonably broad distribution of income and land, along with successful policies of infant industrializers to shield themselves from excessive trade competition from dominant economic powers. Again, such favorable social and political conditions are found in few developing countries today, which also do not have the access to cheap foreign natural resources enjoyed by early industrializers.

Successful Third World development will also require the use of appropriate technologies that foster income distribution and broad-based self-reliant development. Chapter 3 discussed the difference between advanced agricultural methods based on low-input methods and methods based on capital- and energy-intensive chemical agriculture or systems based on genetically engineered plants. In the energy sector, electrification schemes based on large central stations may be useful for supplying large cities. But the same technology is structurally unsuited to reach dispersed townships and villages at reasonable cost and speed.

In summary, it is not clear how international money and technology transfer alone can create the necessary political, social, and structural conditions for successful economic development in Third World nations. Even the much more limited issue of large-scale transfers of efficient, low-pollution technologies to the developing nations raises difficult questions, given that currently most technology transfer occurs through the operation of profit-oriented private multinational firms. However, these questions go beyond the scope of our discussion below, which focuses on the more narrow issue of a global climate convention.

[2]A review and synthesis of the many critiques of the current international economic order, along with analyses and critiques of prevailing development strategies in both socialist and capitalist developing countries, can be found, for example, in Senghaas (1977).

B. INTERNATIONAL PROTOCOLS VERSUS LEADERSHIP BY EXAMPLE

1. The Need for Unilateral Initiatives

The development of a climate convention with effective reduction protocols will require great commitment, leadership, diplomatic skill, and protracted negotiations. Calls for concluding a treaty by 1992 notwithstanding, experience with the Montreal Protocol on chlorofluorocarbons suggests that a 10-year period is not an unrealistic time frame for resolving the many technical issues involved.

But climate-stabilizing action cannot wait that long. The discussion of the previous chapter showed that action to reduce fossil carbon releases must be initiated within the next few years. In view of this timetable, individual nations and groups of nations will need to act even before comprehensive international agreements are concluded.

Here, the major industrial nations, notably the Group of Seven, are under obligation to lead by example. Unilateral or multilateral action by any of these countries could prove effective in creating momentum toward broader participation. Just as recent unilateral actions in the arms control area and in the control of CFCs have overcome negotiating deadlocks, so too could unilateral reductions in fossil fuel use pave the way for broader greenhouse gas control. Action by the wealthiest nations could build indispensable good will needed to conclude effective international treaties and protocols. Unilateral action could also take the form of partnerships between individual industrialized countries and developing countries, which would work toward implementing the global compact on a bilateral level.

Some countries have already taken unilateral steps. For example, the Netherlands has adopted a target to return national fossil carbon releases to 1990 levels by the year 2000, and the Swedish parliament has adopted a pledge to keep fossil carbon emissions in that country below current levels. Though these ceilings are less than what our analysis would indicate is necessary, they are an important signal.

2. Limits to Unilateral Action

The core concern for most countries is whether unilateral cuts in fossil fuel use would negatively affect their competitive international position. This concern gets us back to the discussion in Section A of Chapter 1, in which we emphasized the need for a proper least-cost analysis of the net cost of climate-stabilizing actions. Major carbon dioxide emission reductions could be achieved at negative net economic cost to society, and even larger reductions at negative net economic and social cost, because current institutional and market barriers have led to underinvestments in highly cost-effective energy-efficiency improvements. Depending on the individual circumstances and level of international assistance, even the cost of capital transfer to developing nations could possibly be covered by energy-sector savings.

From this result, unilateral action would seem not only feasible, but could in fact be used by individual nations to strengthen their competitive position. However, there are a number of other potential impediments to unilateral action. One is that large-scale fossil phaseout efforts in some of the major industrialized countries would likely lead to price collapses for fossil fuels and possibly greater consumption of lower-cost energy elsewhere. Also, price-cutting policies by major producers could derail fossil fuel substitution investments before they even get fully under way. Here, special international compensation agreements will be needed with the major fossil fuel exporters (see below).

A second complication arises from international trade. A country may raise fossil fuel prices internally in a revenue-neutral manner, but could face competition from products produced in other countries that did not incorporate global externality costs in the same way in their fuel prices. This competition — and potential shifts of energy-intensive production to other nations in response to that competition — could in principle be fended off by levying an import tax on energy-intensive products from such countries. However, this tariff could be in conflict with the Global Agreement on Trade and Tariffs (GATT).

Logically, it is the GATT agreement that would ultimately have to change: at present, the free-trade regime it purports to maintain is riddled with hidden "subsidies" obtained from uneven use and destruction of global environmental resources in the various parts of the world. The fact that environmental protection standards are not globally applied is yet another reason why notions of free and fair trade tend to be of quite arbitrary character.[3]

Finally, unilateral action by industrialized countries will not be effective if developing countries are not also enabled to use efficient, low-pollution technologies for their economic development. This issue may be the most compelling argument for backing up unilateral action with some kind of international agreement.

[3]In fact, one could argue that GATT should endorse environmental taxes on imports when these would aim at exporters that are delinquent in applying best available technology in protecting globally shared environmental resources such as climate.

But on the way to such an agreement, major initiatives by leading industrialized nations will still be indispensable. To facilitate both early action by these nations and comprehensive participation later, a global treaty might be pursued in stages that encourage some action early in the process, even if participation is limited at first.

Fortunately, the countries with the greatest responsibility to provide leadership are also the ones with the most sophisticated institutional cooperation mechanisms in place. These include the European Economic Community (EEC), the Organization for Economic Cooperation and Development (OECD), and the U.N. Economic Commission for Europe (ECE) countries (which includes the Soviet Union and North America).

3. The IPCC Process: Progress and Challenges

The first formal call for a global climate convention by a broad-based international meeting came in June 1988 at the Toronto Conference on the Changing Atmosphere (Toronto 1988). Government and non-government organizations (NGOs) in various quarters have been developing sketches and drafts of such a convention (see, e.g., Enquête-Kommission 1988; Feiveson et al. 1988). In November 1988, the Intergovernmental Panel on Climate Change (IPCC) was initiated by the United Nations Environment Program (UNEP) and the World Meteorological Organization (WMO). In February 1989, an international meeting of legal and policy experts, sponsored by these organizations and the Canadian government, developed a first draft of a framework for such a climate convention (Ottawa 1989). In March of the same year, 22 nations formally committed themselves to establishing a global climate convention, including five of the seven major OECD countries.

The principal challenge in the IPCC process is to develop reduction protocols with teeth rather than just a general framework with non-binding language. First, an effective disaggregation of the overall problem must be found that avoids losing unnecessary time in overly complex negotiations (Tolba 1989). This concern is reflected in proposals to focus on protocols for individual gases or policy areas, which could be agreed upon one by one.

At the same time, there is considerable interlinkage between the various greenhouse gas emissions, as discussed in earlier chapters. Thus, on a technical level, the challenge is to formulate indices of emission reduction that do justice to the nature of the climate-warming problem. On the political level, the key problem is the adoption of sufficiently strict reduction targets. On the policy level, it will be crucial that the selected abatement responses are in harmony with broader sustainable–developmental considerations.

This task is further complicated by the international equity issue. The Third World contribution to the overall greenhouse effect is already significant and growing (see Chapter 5) and must be addressed right away. One way of doing so is to build into each protocol a compensatory mechanism that creates economic incentives for low-pollution investments while providing funding for corrective measures in the capital-strapped developing countries.

In this concluding chapter, we highlight how the scientific and policy analyses presented in this report might be incorporated in the current IPCC process. Given the great complexity of an international climate convention, the following comments and suggestions should be understood as selective contributions, and not as an attempt to present a comprehensive protocol framework.

C. THE CLIMATE CONVENTION

1. Basic Agreements

The overall framework for a climate convention has been sketched in Ottawa (1989). The focus here is on its substantial content, which might look as follows:

1. The nations of the world agree to a warming limit: anthropogenic trace gas emissions will be reduced sufficiently to limit global warming to the temperature regime in which humanity and most of the world's biological gene pool evolved, that is, a 2–2.5°C ceiling. Furthermore, the rate of climate warming is to be reduced to match the capacity of forest ecosystems to adapt to warming, or to no more than about 0.1°C per decade.

As shown in this book, adoption of these limits on warming probably minimizes climate risks to levels we will simply have to tolerate. Of course, higher warming ceilings could be chosen as a basis for a convention. But, as discussed in Chapters 1 and 2, even this minimum feasible limit on warming poses major climate risks. A significantly higher warming limit would make climate stabilization progressively more elusive while disproportionately escalating the risks of devastating impacts on society.

2. For planning purposes, the signatories of the climate convention take the period between now and 2100 as a reference point. They agree to reverse the current rising trend of radiative forcing and equivalent carbon dioxide concentrations in time to achieve close to level global surface temperatures by the end of the next century.

3. These warming limits are given explicit interpretation in terms of maximum allowable equivalent CO_2 concentrations. This interpretation is based on the risk-minimizing assumption of upper-range climate sensitivities in excess of $4°C$ for a doubling of CO_2.

4. These risk-minimizing specifications are further translated into a package of explicit protocols for individual trace gases and policy areas, notably
   - A cumulative global fossil carbon budget.
   - A target for the long-term level of biotic carbon storage relative to the present level.
   - A near-term phaseout schedule for chlorofluorocarbons.
   - The global implementation of strict controls on acid rain and other classical air pollutants.

Based on our analysis in this report, these global targets should be as follows:

- The cumulative fossil carbon budget for the 1985–2100 period should be about 300 billion tons.
- Biotic carbon storage should be returned to present levels.
- Chlorofluorocarbons (CFCs) should be phased out by the turn of the century.
- Classical air pollutant emission rates should be reduced by percentages equivalent to the full application of best available technologies worldwide.

The reasons for treating fossil and biotic carbon releases in separate protocols are further discussed below.

5. The above goals and requirements should be subject to periodic review to incorporate new scientific evidence.

6. The convention should focus on achieving agreement among the 15–20 largest trace-gas-emitting industrialized and developing nations. For equity reasons, the convention should also disaggre-

gate the world's nations into developing and industrialized countries, based on the established categorizations of international agencies. As further discussed below, protocols on emission reductions should be formulated on the basis of separate requirements for each of the two groups of countries.

Two other subgroups of countries should be given special attention: the industrialized countries of Eastern Europe and the Soviet Union, which are currently undergoing a difficult period of internal restructuring, and the oil-exporting and coal-rich developing countries for whom climate stabilization will mean a significant reappraisal of their national resource assets.

2. Should Fossil and Biotic Carbon Emissions Be Linked?

In developing a protocol for carbon dioxide emissions, a key issue is whether and how to link fossil carbon emissions with those from deforestation and soil erosion. On the surface, linking the two sources and allowing trade-offs between fossil reductions and net reforestation efforts[4] appears to be an elegant and sensible solution. Upon closer examination, there are reasons not to pursue such an offset approach at the global level at this time:

- Carbon fixing through tree planting and fossil fuel conservation differ from each other in fundamental ways and are not equivalent. Reforestation and forest preservation provide only temporary emission reductions until biomass levels in replanted areas have steadied. By contrast, fossil carbon substitutions can provide cumulative emission reductions as time goes on. Also, carbon sequestration in forests and trees is in principle reversible, while fossil emissions reductions are not (see Chapter 3).
- Compared to fossil fuel consumption, net reforestation and deforestation rates are hard to measure and verify. As discussed in Chapter 3, the current rate of deforestation in various parts of the world is highly uncertain. While such uncertainties could be addressed within a deforestation/forest and soil conservation protocol, the linkage of highly uncertain biotic CO_2 emissions with the much better understood fossil fuel emission data is inherently unsatisfying.

[4]Net reforestation means the net effect of reduced deforestation, afforestation, and reforestation; see Sections D–F in Chapter 3.

- The offset approach could reinforce the misguided notion that any kind of tree planting is environmentally and socially benign and sustainable. As discussed in Chapter 3, forestry projects can themselves aggravate rural poverty. While offset linkages would make the capital-rich energy supply sector interested in financing forestry projects, these new players could also inadvertantly aggravate the detrimental impacts of many commercial forestry projects on the Third World poor. This would not only have negative social consequences, but could ultimately foil the environmental purpose of these projects.
- Specifically, successful reforestation and forest preservation requires a participatory approach at the grassroots level and, in many cases, far-reaching land-use reforms. Altering fossil fuel consumption is an inherently more centralized task than changing land-use patterns and biomass economies. It could be difficult to coordinate the necessary national policy-making process in developing countries with the emission-offset agenda of energy supply companies.
- More broadly, linkage between the fossil fuel area and the reforestation/forest and soil preservation area could easily aggravate problems of interference by industrialized nations and their powerful commercial organization in the national planning of other nations.
- There are also complexities in the equity area. For instance, the much higher per capita beef consumption in the industrialized countries is based, in part, on the use of large tracts of land for agricultural feedstock production that once were forested, and could again be forested if beef consumption were lower (see Chapter 3 and the discussion below).

For these reasons, we recommend that separate protocols be formulated, at least initially, for the fossil carbon and biotic carbon (mainly deforestation) emissions. Once separate protocols have been established in each area, experience has been gained in implementing them, and the data situation on biotic carbon storage has been improved, linkage could still be introduced if advantageous.

3. Global Climate Protection Fund

Part of the framework should be an international fund that supports climate-stabilizing investment and research and development projects and policies, especially in the developing world. This idea has been put forth in several proposals for a global climate convention. The Ottawa meeting of legal and policy experts adopted the idea of a fund ("world climate trust fund") in February of 1989. Such a fund, and parallel bilateral and multilateral assistance programs, should be structured to reflect the north-south compact on sustainable development. It would therefore differ from conventional development assistance mechanisms in fundamental ways.

Mission. A principal use of the fund would be to pay capital-strapped developing countries for the extra cost of installing and/or retrofitting the most energy- and pollutant-efficient technology instead of standard technology. A further purpose would be to support policies and projects aimed at reforestation and the widespread introduction of innovative agroforestry, tree cropping, and other low-input farming techniques. Both the transfer of currently available technology and research and development and commercialization efforts would be financed, and both governments and nongovernment organizations would be eligible for support. Besides providing capital for preventative action, the fund would also contribute to adaptation costs and emergency relief.

Aside from financing such efforts, a section of this fund may also have to be used to compensate the minority of fossil-rich developing countries for leaving fossil fuel resources in the ground. The possible need for such compensation has mostly been overlooked so far, partly because most discussions to date were not based on the strict climatic risk limits used in this report. As shown in Chapter 4, climate stabilization will require that fossil fuel consumption be limited to only a portion of available low-cost coal, oil, and conventional gas resources.

This latter use of funds would be similar to the agricultural subsidies paid to farmers in the United States and elsewhere to retire farmland and limit production. Disbursements should be primarily aimed at those countries that are heavily dependent on their reserves of domestic fossil resources for their economic development and whose cumulative per capita emissions to data are comparatively small, such as coal-owning China and India, and the OPEC countries. In order to assure economic efficiency, lockup agreements could be purchased through auctions in which the fossil reserves with the highest production costs would be offered first.

Funding Sources. Payments into a global climate protection fund and capital for other dedicated bilateral and multilateral environmental funds could be raised in a number of ways, including

- Industrialized countries could charge themselves user fees for climate-changing and environmentally damaging activities, such as fossil fuel consumption and energy use at large, beef consumption, CFC production, and so on.
- ICs could provide debt relief for developing nations, which could take the form of
  - Combinations of debt reduction and conversion of debt payments into fund contributions for environmental programs.
  - Debt-for-nature swaps in which discounted debt is purchased and converted into conservation bonds that would generate local funding for environmentally benign income-generating programs and natural resource protection projects.

The latter approach could be particularly appealing in the area of tropical deforestation, since two-thirds of global forest losses and half of Third World debt occur in the same 14 developing countries (Speth 1989). However, the concept of debt-for-nature swaps should be treated with caution. If such swaps remove current sources of foreign-exchange earnings without accompanying deliberate investments in alternative income-generating ventures, they may neither solve the debt problem nor remove the poverty-related driving forces of environmental degradation.

A global fund also raises the issue of formulas or guidelines for determining appropriate contribution levels from individual ICs. Contributions from individual industrialized countries could be made proportional to indices based on past and current emissions, as illustrated in Section E of Chapter 5. Further suggestions are presented below. Though the suggested formulas are based on a rationale and calculus that measures international inequity, these proposals do not have to be used as rigid quotas. Instead, they could be used for setting upper- and lower-limit guidelines in a sliding-scale approach that takes into account each country's ability and willingness to pay.

Ecodevelopmental Conditionality. Unlike existing international assistance and lending agencies, the climate protection fund should have an explicit mandate to further sustainable development. In operating this international fund, it will be of vital importance to subject all funded projects to a careful environmental and social review process. The history of international lending and development assistance projects is replete with examples that inadvertently created negative environmental and social impacts, often far in excess of project benefits.

Projects aimed at climate stabilization cannot automatically be assumed to be free from such adverse impacts. Each project or policy measure should therefore fulfill an ecodevelopmental conditionality: Specifically, projects should be screened to avoid causing or magnifying the following inadvertent impacts:

- Inefficient use of capital due to an overly supply-oriented approach based on large-scale, centralized technologies, and projects that overlook cheaper options, such as increased demand-side efficiency and more dispersed, medium-scale or modular systems.
- Income or land ownership concentration and increased poverty among local populations, for example, in the form of displacement of tribal subsistence economies, or through the commercialization of previously free environmental goods without commensurate income generation in the local economy.
- Major environmental impacts in areas unrelated to climate change.
- Risks of major catastrophic accidents.
- Risks of nuclear weapons proliferation, including increased risk of use of nuclear and other technologies for terrorist activities.

Mandatory, comparative, least-cost analyses of alternative technology options, and review by independent bodies should be part of the operating guidelines for the fund. Finally, disbursement of funds should be made conditional on the safeguarding of the civil rights of citizen groups engaged in local social and environmental action.

Control of the Fund. Unlike current development assistance agencies that are totally controlled by the industrialized "donor" countries, a climate protection fund would need to be administered with major participation by the developing countries. Such Third World representation is appropriate, since the capital transfers for the purposes of global environmental maintenance and repair are for a joint venture, and not welfare payments to the poor.

Also, it should be evident that even if the nations of the world band together to save the world's climate, this will not automatically do away with the deep division among rich and poor nations. Adequate Third World participation is needed to prevent forms of "environmental colonialism" in which the climate issue is inadvertently or deliberately used to reinforce traditional agendas that are in conflict with the north–south compact.

To ensure balanced fund operation, representatives from nongovernmental organizations (NGOs) should also have an oversight role. NGOs have proven effective leaders in environmental action because of their scientific and social expertise; both will be indispensable in the future.

D. FOSSIL FUEL PROTOCOL

1. Cumulative Fossil Carbon Budgets and Reduction Milestones: The Basic Bargain

The basic arrangement that would guide this protocol is as follows:

- Industrialized countries and developing countries will commit themselves to base their energy planning on roughly equal shares of a global fossil carbon budget of 300 btC, and adopt reduction milestones and peak emission caps along the line of the analysis in Chapter 6.
- ICs will provide the capital and technology assistance to allow DCs to buy and produce state-of-the-art efficiency technologies and renewables-based energy systems.
- ICs and DC fossil fuel exporters will make special agreements to stabilize global fossil fuel prices and lock up the more expensive fossil fuel reserves.

The concentration limit/emission budget/reduction milestone approach is suggested as the most sensible basis for a fossil carbon protocol. The advantages of a carbon budget/reduction milestone approach over other approaches have been explored in Chapters 1 and 5.

Our analysis in Chapters 5 and 6 suggests that for a 300-btC global budget, a 50–50 split would be a fair and workable guideline. The corresponding reduction milestones for industrialized and developing countries were explored in Chapter 6. Chapters 5 and 6 also illustrated possible formulas for deriving fossil carbon budgets for individual countries. Since actual fossil fuel needs may vary from projected requirements even when a phaseout plan is meticulously implemented, and much more so in a world of unexpected events, these carbon budgets would be used as a benchmark for compensatory financial arrangements among nations.

2. Possible Mechanisms for Financing Third World Energy Investments

Chapter 5 showed that a strictly equitable sharing of the global fossil carbon budget is not feasible on logistic grounds. However, the person-year equity approach could be made the basis of the initial capitalization of a climate-protection fund, as described in Chapter 5. The same discussion showed that a modest, 10 percent surcharge on fossil fuels would mobilize several tens of billions of dollars—a sum comparable to all current bilateral and multilateral development assistance combined.

The mechanism proposed in Chapter 5 would provide not only an equitable guideline for financing internationally sponsored corrective measures but also an economic incentive for reductions in fossil fuel use at home.[5] Moreover, the largest amounts of capital raised through this mechanism would be available in the initial two to four decades, when industrialized country emissions will be most out of line with their fair share, and the need for massive investments in developing countries will be most pressing.

Again, not all ICs are currently able to take on major capital transfer burdens equally well, and individual circumstances will need to be taken into account. Fortunately, recent polls in the wealthiest industrial nations have shown an increasing willingness to pay for global environmental repairs, which could facilitate pragmatic compromises.

3. The Need for Detailed National Least-Cost Reduction Plans

One of the logical components of a fossil carbon protocol would be the periodic filing of national emission reduction plans with an appropriate coordinating agency. Such plans could be one of several tools for monitoring the progress of implementation. The usefulness of such reduction plans will greatly depend on developing more consistent energy analysis practices. The key methodological and data source issues are

- The use of appropriate least-cost principles, including consideration of noneconomic costs and risks, in formulating resource plans.

[5]Such energy taxes by themselves are insufficient to bring about the necessary reductions. Regulatory standards, revenue-neutral mechanisms that tie taxes to incentives and subsidies for fossil-substituting investments, and other least-cost planning policies will need to be implemented as well.

- The integration of efficiency (in particular, demand-side) investments into energy planning.
- The availability and proper generation of detailed end-use data and baseline statistics on existing efficiency levels needed for quantifying demand-side resources.
- Consistent cost evaluations of alternative technology options.

These issues need to be much better resolved in order to come to agreements on what levels of carbon reductions are economically feasible. A number of international cooperation efforts, both by non-government organizations (NGOs) and governmental bodies, have been started to address these issues.

E. PROTOCOL ON REFORESTATION AND AGRICULTURAL LAND USE

1. The Basic Bargain

The guiding principle of a deforestation protocol should be as follows:

- The nations of the world will take action to maintain and restore global carbon storage in land biota to mid-1980s levels.

This goal flows out of the analysis of Chapter 3, and underlies the fossil carbon budget derivation in Chapter 4.

Given the analysis in Chapter 3, a protocol on deforestation and reforestation could be linked to the per capita area of afforestable cropland and pasture under production in developing and industrialized countries. In Chapter 3, it was pointed out that the most difficult aspect of tropical deforestation is the real increase in demand for agricultural land. Even after all the other contributing factors to deforestation have been brought under control, there remains the challenge to increase food production without commensurate forest clearing.

We also explored how tree cropping and productivity increases in Third World cropland agriculture could be used to limit the need for new cropland and pasture despite the anticipated doubling of population. As noted, a significant portion of Third World cropland and pasture is devoted to food production for consumption in the industrialized countries. At the same time, our analysis showed that on a per capita basis, industrialized countries use four times as much nondryland cropland and pasture as developing countries. Pastures and croplands outside dryland areas are the prime candidates for afforestation. Moreover, much of the disproportionately large use of such prime land for agriculture in the industrialized countries is due to the five- to sixfold higher per capita animal protein consumption, particularly beef consumption, in these countries.

A deforestation protocol could account for these factors through the following agreements:

- Developing nations set themselves a target of limiting the growth in croplands and pastures from forest clearing to a certain fraction of population growth.
- Industrialized countries agree to provide major funding and technology assistance to increase Third World agricultural productivity, notably through advanced low-input techniques.
- Industrialized countries also undertake compensating afforestation activities at home that offset unavoidable losses of Third World forests.

These elements are briefly discussed below.

2. A Special Forest Clearing Allowance for Developing Countries

Given the expected doubling of populations in developing countries over the next few decades, there is an obvious need to bring more agricultural land into production. A major portion of this new agricultural land will need to come from currently forested land. In Chapter 3, it was shown that productivity in many Third World agricultural systems can be increased by up to a factor of 2–4 even without introducing fertilizer- and pesticide-intensive chemical agriculture.

This means that a major portion of the additional demand from population growth could in principle be met by productivity increases. We also pointed to the option of using tree-crop agriculture and agroforestry schemes to reduce the amount of tree carbon losses due to food production. However, prerequisites of such improvements are development assistance programs and land and policy reforms that could free rural populations from the short-term survival pressures that prevent them from adopting more advanced low-input methods now.

To give a quantitative illustration of a forest clearing allowance based on a population growth/agricultural land growth formula, suppose developing countries were to determine that with adequate capital assistance to finance extension services and land reforms, they could limit the growth rate of agricultural land to one-quarter of population growth. For the high-growth period between 1985 and 2025, this limit would translate into a cumulative forest-clearing allowance of about 130 million hectares, or 7 percent of currently existing closed and open Third World forestlands.[6] Limiting Third World forest clearing to this allowance would still be a great improvement over present trends: permissible clearing would be only about one-fourth of the loss of forests from continuing current net deforestation rates in the tropics.[7]

Of course, this forest-clearing allowance would have to be applied in a sensible manner that does not pit food production and human survival against forest preservation. It can only be implemented to the extent that the necessary productivity improvements and shifts to agroforestry and tree cropping are indeed realized.

3. Compensating Carbon Storage in the Land Biota of Industrialized Countries

To maintain the goal of holding biotic carbon storage steady on a global level, industrialized countries would have to afforest an area of about 130 million ha, equivalent to taking about 20 percent of their (nondryland) cropland and pasture out of production.[8] ICs would be faced with the challenge of reducing agricultural land requirements at a time when they also need to convert from chemical, fertilizer-intensive agriculture to sustainable organic practices. As the discussion in Chapter 3 showed, this transition is likely to limit offsetting productivity gains in the near-to-medium term. But properly conceived, afforestation could become part of a strategy for regenerating IC farmlands.

[6]Third World rainfed cropland areas and pastures were about 700 million hectares in 1982–1984 (see Table 3.11). Assuming the U.N. medium population growth projection of about 1.7 percent per year for the developing world between 1985 and 2025, agricultural cropland and pasture would be targeted to grow at a rate of about $1.7 \div 4 = 0.43$ percent per year, or 19 percent over 40 years. This is equivalent to $0.19 \times 700 = 133$ million ha, or about 7 percent of the roughly 2 billion ha of open and closed forests in the Third World (see Table 3.11).

[7]If current deforestation and degradation rates of about 0.6 percent per year (see Section D in Chapter 3) were to continue unabated over 40 years, some 27 percent of Third World forests would be lost.

[8]For the purpose of this exploratory discussion, we assume a rough carbon equivalence between the average cleared hectare in the developing world and the average afforested hectare in the industrialized regions.

Beyond that, a more balanced diet with a moderate shift from beef consumption to other forms of animal protein or to nonanimal protein sources could provide room for the necessary adjustment.

4. Possible Mechanisms for Financing Third World Agricultural Improvements

Dedicated funds for agriculture-related capital and technology transfer could be raised in one of several ways:

- A reforestation tax on beef consumption in the industrialized countries.
- Improved prices and terms of trade for agricultural export commodities of developing countries, tied to switching some land now devoted to export crop production to domestic food production.
- An IC consumption surcharge on Third World cash crop commodities, such as coffee, sugar, and other tropical products, to be recycled into a dedicated fund for Third World agriculture and forestry projects.

The beef consumption tax would be similar to a carbon tax on fossil fuels. The tax would provide an incentive to limit or reduce beef consumption and thus the need for land-intensive feedgrain production. It would also reduce methane releases from livestock. In many industrialized countries, the net impacts of such a policy could be less severe than one might expect: reductions in meat and dairy production could solve overproduction problems and save large amounts of farm subsidies. In fact, part of such a tax could be used to support IC farmers in their transition to organic farming and to give them a broader role as "ecotope stewards."

Higher prices for those Third World agricultural export commodities that do not compete with IC domestic production, that is, the tropical products that cannot be grown in temperate zones, could be an effective way of providing growth stimuli in the short term. IC consumers have weathered significant price oscillations for these commodities in the past, and have benefited from the collapse of many commodity prices over the last 10–15 years. The partial conversion of export cropland to cropland for domestic food production could be used to prevent overproduction in response to higher prices.

Guidelines for relative IC contributions to such an international reforestation/agriculture fund could be based on agricultural land-use patterns. One possible formula would be to calculate the difference between the rainfed cropland and pasture actually under production and the land needed to produce a balanced per capita diet at reference productivity levels.

To illustrate this approach, we might consider meat consumption, which dominates land requirements for agriculture in the industrialized regions. Assume that for purposes of establishing contribution guidelines, a balanced diet is defined per international agreement as including twice the World Health Organization (WHO) minimum daily requirement of animal protein of 30 grams per capita per day, or 60 g/cap-d. On average, developing countries consume only about 38 g/cap-d, while the figure for industrialized countries is 210 g/cap-d (see Section C in Chapter 3).

Contributions to the reforestation/agriculture account could be made proportional to the difference between actual per capita land requirements to produce the animal protein actually consumed, and the land required to produce 60 g/cap-d, using, for example, average IC and DC productivity as reference standards (again, the 60-g figure is purely illustrative).[9] Alternatively, animal protein consumption, weighted by the feedgrain inputs for each type of meat, could be made the basis for calculating financial contributions.

5. Forest-Cover Monitoring and Standardized Carbon Storage Factors

The first component of a deforestation protocol should be efforts to greatly improve the knowledge about current rates of deforestation, and the associated carbon releases. Also, a sufficiently detailed and accurate system of assigning carbon-fixing rates and carbon-storage levels to different land uses and replanting schemes should be developed. This would allow the standardized determination of each country's or region's net deforestation and reforestation rates. Statistics on degraded forestland should be improved and factored into this rate.

---

[9]There are important complexities in calculating this land requirement. First, import–export flows must be taken into account. Second, a much greater portion of IC meat consumption is based on beef. The beef in turn is grainfed to a much higher degree than in the Third World. On the other hand, the higher productivity with which feedgrains are produced in ICs provides some offset. However, overall, the differences in animal protein consumption are in proportion to per capita agricultural land uses in the two groups of countries.

F. PROTOCOL ON CHLOROFLUOROCARBONS

The Montreal Protocol is currently being revised, with the goal of phasing out chemicals that deplete stratospheric ozone as early as by the year 2000. As mentioned in Chapter 3, the ozone problem imposes even more stringent limits on the use of these chemicals than does the greenhouse effect. As a result, the revised Montreal Protocol should simply be made part of the climate convention. Of course, substances with low ozone depletion potentials must be screened for their greenhouse impacts.

G. PROTOCOL ON OTHER TRACE GASES

Among others, the following subagreements suggest themselves:

- An expanded protocol on the control of acid rain emissions, carbon monoxide, and other classic air pollutants. What is needed, in essence, is the globalization of tough air pollution legislation for vehicles and stationary sources. Existing acid rain control agreements and national legislations on automobile emissions can be used as a starting point, but need to be tightened. Pollution control in developing countries should be financed in a manner similar to efficiency improvements. Moreover, efficiency improvements could themselves be used to reduce pollution while achieving fossil carbon benefits as well.
- Agreements to limit or eliminate the flaring off of natural gas and other sources of fossil hydrocarbons; to introduce recycling requirements in place of landfills as the principal waste management strategy; and to limit the methane production from livestock in industrialized countries, for example, by implementing the beef consumption tax proposed above.
- A milestone schedule for reducing the use of chemical fertilizers in industrialized countries.

As in the case of the other protocols, these agreements will need to be accompanied by research to improve current knowledge on the nature of the sources and chemical cycles involved.

H. CONCLUSION

The above agenda raises political and implementation questions that seem overwhelmingly difficult, given past experience with international cooperation. Nobody has yet been able to figure out how the political will can be mobilized to undertake such unprecedented global action.

But like any challenge in human life, the global climate threat can also be looked at as an opportunity. Humanity is being spurned to undertake with vigor the transition to a highly resource-efficient, equitable, and sustainable civilization. The world has today all the necessary technical and research tools to undertake this transition. What is needed now is a clear action plan for using them, and the enlightenment to implement it with alacrity.

Surely, the required profound changes in economic development and human resource consumption will be slow to materialize so long as their economic costs are seen as overwhelming. Specifically, there is a mistaken widespread belief that the costs of rapidly phasing out fossil fuels would be prohibitive, and if begun unilaterally, could harm the competitive position of countries doing so.

Energy strategies for climate stabilization and the net economic cost of warming prevention in the energy sector must be investigated. The following three questions arise:

1. How much lower could fossil fuel consumption be if all efficiency improvements and other carbon-emission reducing options were implemented that are economically viable at present energy prices, but whose implementation is being blocked on account of market barriers and market-distorting government policies?

2. How would this least-cost level of fossil carbon emissions change if the externality costs and impacts of each fossil and nonfossil energy source were properly accounted for?

3. How would the levels of fossil fuel consumption identified under items 1 and 2 compare to climate-stabilizing reduction targets? And what, if any, would be the incremental cost to society of moving from the level of fossil fuel consumption identified under the above items to climatically sound levels?

These fundamental questions have remained largely unaddressed and unanswered by officially sponsored energy strategy and policy research. To support a global plan for climate stabilization, they should become the focus of significant and ongoing national energy policy projects throughout the world.

REFERENCES

Enquête-Kommission (1988). *Erster Zwischenbericht, Enquete-Kommission Vorsorge zum Schutz der Erdatmosphaere of the FRG Parliament,* Deutscher Bundestag, Bonn, Nov. 2.

Feiveson, H. A., et al. (1988), *Princeton Protocol on Factors That Contribute to Global Warming,* Woodrow Wilson School of Public and International Affairs, Princeton University, Princeton, NJ, Dec. 15.

Ottawa (1989), "Statement of the Meeting of Legal and Policy Experts," Protection of the Atmosphere: International Meeting of Legal and Policy Experts, Ottawa, Feb. 20–22.

Senghaas, D. (1977), *The International Economic Order and the Politics of Development* (in German), Suhrkamp, Frankfurt.

Senghaas, D. (1982), *Lessons of the European Development Experience* (in German), Suhrkamp, Frankfurt.

Speth, J. G. (1989), *Coming to Terms: Toward a North–South Bargain for the Environment,* World Resources Institute, Washington, D.C.

Tolba, M. K. (1989), "A Strategy for Success," Keynote address to a Meeting of Legal and Policy Experts on the Protection of the Atmosphere, Ottawa, Feb. 20–22, UNEP.

Toronto (1988), "The Changing Atmosphere: Implications for Global Security," Environment Canada, Conference statement, Toronto, June.

Summary and Conclusions

The study presented in this book investigates global climate stabilization targets for energy policy and planning. It was performed by an international team of experts for the Dutch Ministry of Environment. The report complements other major policy studies on global climate in significant ways:

- While previous analyses provided menus of "warming fates" based on projections of trace gas emission trends and policies, this study examines required actions on the basis of explicit risk-minimizing limits on the rate and magnitude of global warming (Warming limit/Emission budget/Reduction milestone Method, or WERM).
- In assessing required emission curtailments, the present study uses modeling calculations of both the equilibrium response and the transient response of the climate system.
- The study provides an assessment and ranking of important technical options and strategies for reducing fossil carbon emissions in terms of their net economic cost.
- The study explicitly addresses how industrialized and developing nations might share responsibility for climate stabilization on the basis of international and intergenerational equity as well as practical feasibility.

OVERVIEW

To provide the necessary context for our climate-stabilizing energy policies we start from an analysis of both energy and non-energy related driving forces of the greenhouse effect. We develop a budget for future carbon dioxide releases that would meet specified limits on the risks of human-induced climate change. We examine how carbon dioxide and other releases will need to be curtailed in the future, and how individual regions and countries could do their fair share in dealing with this global problem. The final chapters provide guidelines for the formulation of *an international convention on climate stabilization and sustainable development.*

PRINCIPAL FINDINGS

The study presented in this book finds that:

- The centerpiece of an international agreement to protect the world's climate should be a global budget for cumulative fossil carbon releases between now and 2100.
- This budget should be based on a policy of risk minimization, including explicit limits on the rate of global warming and on the absolute magnitude of warming.
- This fossil carbon budget should take into account impacts of trace gases other than carbon dioxide, and carbon dioxide releases from sources other than fossil fuel combustion. The implied emission reductions for these other trace gases should be made explicit.
- According to the risk assessment used in this study, the average rate of warming should be limited to about 0.1°C per decade. Currently, a warming commitment of 0.2–0.5°C per decade is being added. The absolute warming relative to 1850 should not exceed 2.5°C. Limits on radiative forcing and maximum equivalent-CO_2 concentrations should be calculated using the higher end of the current range of climate sensitivities.
- A climate stabilizing fossil carbon budget for the period from 1985 to 2100 would then be about 300 btC (billion tons of carbon, or gigatons), equivalent to about 55 years of fossil releases at current rates. This budget implicitly assumes that
    Chlorofluorocarbons (CFCs) will be phased out by the end of the twentieth century.

Carbon storage in forests and soils will be returned to mid-1980s levels through several decades of afforestation in both temperate and tropical regions, combined with curtailment of tropical deforestation. Increased food production will be achieved mainly through productivity increases rather than land-clearing, and carbon-storing tree cropping and agroforestry systems will be widely adopted over time.

Overall methane emissions will be somewhat reduced by limiting agricultural emissions and by decreasing emissions from fossil fuel production and use.

The growth of atmospheric nitrous oxide concentrations will be slowed.

- Such a fossil carbon budget implies the following emission reduction milestones:

   Global releases should have returned to present (ca. 1985) levels by about 2005, or within 15 years.

   A 20 percent reduction below 1985 levels should be achieved by about 2015, or within 25 years.

   A 50 percent reduction should be achieved by about 2030, or within 40 years.

   A 75 percent reduction should be achieved by about 2050, or within 60 years.

- Relative to populations, cumulative historic fossil carbon emissions from industrialized countries are more than eleven times as high as those from developing countries. If the remaining global fossil carbon budget were shared according to strict person-year equity including historic emissions, industrialized countries would have no emission rights left.

- A reasonable compromise between international equity and practical feasibility would be to allocate 150 btC each to industrialized and developing countries. This allocation would imply the following targets:

- ICs should achieve a 20 percent reduction in fossil carbon releases by 2005, a 50 percent reduction by about 2015, and a 75 percent reduction by about 2030.

- Developing countries as a group should aim to stabilize their fossil fuel consumption early in the next century, and limit increases in the meantime to about 50-100 percent above 1985 levels.

- By about 2030, DC fossil carbon releases should have returned to approximately present levels.

- Put in terms of the fossil carbon intensity of gross domestic product (C/gdp ratio), the targets for the near to medium term (late 1990s to about 2010) would read about as follows:
- Assuming a gdp growth of 2.5–3%/yr, industrialized countries as a group would need to achieve reductions in C/gdp ratios of about 5%/yr by the turn of the century, or about twice the 1979–1986 rate of change.
- Assuming a gdp growth of about 4%/yr, developing countries as a group would need to overcome their lockstep C/gdp pattern beginning in the 1990s. They would need to achieve reductions in the C/gdp ratio of about 3%/yr shortly after the turn of the century in order to sufficiently moderate the unavoidable near-term growth in their absolute fossil carbon releases. This reduction rate would be similar to those realized during 1979–1986 in developing countries with the most proactive energy policies.
- In later years, efficiency improvements, cogeneration, and renewables would need to be introduced at significant speed. Additional carbon releases from fossil-supplied growth in energy services would have to be less than the decrease in carbon releases obtained from retrofitting and replacing inefficient existing plant.
- Implementing these targets will require both unilateral initiatives and cooperative international action:
- Because the urgency of early reductions in fossil carbon releases is great, individual nations should not wait for international agreements but take leadership through prompt action at home. Such leadership is particularly important from the largest and most affluent industrialized countries.
- Contrary to common belief, action by individual countries could bring them economic benefits and increased international competitiveness. This applies also to developing countries since energy-using equipment there is even more inefficient than in industrialized countries.
- Not all nations are equally well equipped to implement the required reductions. Major capital and technology transfers to developing countries will be needed. Some support may also be required by the centrally planned industrialized economies.
- Successful climate stabilization will ultimately depend on a broader compact among industrialized and developing countries to reform the world economic and ecological order and to promote sustainable development worldwide.

OUTLINES OF A GLOBAL COMPACT ON CLIMATE STABILIZATION AND SUSTAINABLE DEVELOPMENT

Basic Propositions

1. All national planning and development must be realigned to observe global environmental constraints.
2. The impoverishment of the developing countries (DCs) is in conflict with this goal.
3. So is the inefficient (and therefore excessive) resource consumption of the industrialized countries (ICs).
4. Climate stabilization must be seen in this broader global development context. To succeed, it must be approached as a joint venture of ICs and DCs to achieve global sustainable development.

This joint venture could be structured as follows:

- ICs will drastically reduce their disproportionate emissions and direct and indirect consumption of global environmental resources.
- DCs will find alternative development paths that avoid the repetition and magnification of IC impacts.
- ICs will provide capital, technology research, development, and transfer, and debt and trade relief to help restart Third World development in a sustainable direction.
- DCs will engage in internal policy reforms and programs to overcome their dual economies, mobilize local communities, engage in land reforms and major population control programs, and build up their own technological research and development capacities.

The capital for this agenda could be found from the following sources:

- reduced military spending,
- major capital-saving efficiency improvements in the IC energy systems,
- other labor and capital productivity gains, and
- economy-of-scale effects of global markets for highly efficient end-use and new and renewable supply technologies.

PROPOSED BENCHMARKS FOR A
GLOBAL CLIMATE CONVENTION

A. Agreement on Minimizing Global Warming Risk

1. The reference planning period for global climate stabilization is the period from now to the year 2100.
2. The nations of the world agree to undertake policies and programs to limit climate warming to the following ceilings:
   - The absolute limit for sustained global surface warming in the period until 2100 and beyond is to be 2.5°C relative to 1850, or ca. 2°C relative to now.
   - The maximum average rate of warming is to be about 1°C per century.
   - The rate of accumulation of warming committments toward the end of the next century will be reduced to much less than 0.1°C per decade.
3. In determining emission reduction targets, these warming limits are interpreted in a risk minimizing manner by assuming climate sensitivities of at least 4–4.5°C for a doubling of carbon dioxide concentrations.

B. Specific Limits for Individual Trace Gases

1. The nations of the world agree to base energy strategies and policies on a budget of 300 billion tons for cumulative fossil carbon releases between now and the year 2100.
2. The nations of the world agree to return carbon storage in land biota to present levels and maintain these levels over time, thus eliminating all net biotic carbon releases to the atmosphere.
3. The nations of the world agree to phase out the production of chlorofluorocarbons and related chemicals by the turn of the century.
4. The nations of the world agree to limit emissions of chemically active trace gases through cooperative programs to implement best available efficiency and pollution control technologies worldwide.

Appendix 1

Empirical Expressions for Calculating Trace-Gas Concentrations

Empirical Expressions for Calculating Trace-Gas Concentrations (C) for Specific Years (y) and the Equivalent CO_2 Concentration ($C(CO_2^*)$) for the Effect of All Gases.[a]

Period	Formula
	CO_2 (ppm)
1790–1870	$C = 279.8 + 0.126 \, (y - 1790)$
1870–1925	$C = 288.5 + 0.00276 \, (y - 1847.2)^2$
1925–1958	$C = 300 + 0.00232 \, (y - 1877.7)^2$
1958–1985	$C = 311 + 0.0206 \, (y - 1944)^2$
≥1985	$C = 291.4 + 0.0135 \, (y - 1921.6)^2$
	CH_4 (ppb)
1765–1951	$C = 790 + 0.0101 \, (y - 1765)^2$
≥1951	$C = 1643 + 14.81 \, (y - 1985)$
	N_2O (ppb)
1900–1975	$C = 280 + 0.000042 \, (y - 1900)^3$
≥1975	$C = 258.5 + 0.0068 \, (y - 1899.1)^2$
	CFC11 (ppb)
1950–1985	$C = 0.00018 \, (y - 1950)^2$
≥1985	$C = -0.0858 + 0.00013 \, (y - 1936.5)^2$
	CFC12 (ppb)
	$C = 1.7 \, \text{CFC11}$
	Equivalent CO_2 concentration
	$C(CO_2^*) = 287.4 \exp{(\Delta Q(\text{total})/6.333)}$
	where $\Delta Q(\text{total})$ is obtained from Appendix 2

Source: After Wigley (1987).

[a]The projections beyond 1985 are fitted to the growth rates given by Ramanathan et al. (1985).

Appendix 2

Estimation of Changes in Radiative Forcing due to Trace-Gas Changes and the Resulting Changes in Global Average Equilibrium Surface Temperature

The change in surface warming is determined by the radiative effect of the trace gases on the radiative heating of the entire surface–troposphere system (Ramanathan 1982), that is,

$$\Delta Q = \Delta Q(A) + \Delta Q(s)$$
$$\Delta Q(A) = \Delta F{\downarrow}(h) - \Delta F{\uparrow}(h) - \Delta F{\uparrow}(O)$$
$$\Delta Q(s) = \Delta F{\downarrow}(O)$$
$$\Delta Q = \Delta F{\downarrow}(h) - \Delta F{\uparrow}(h)$$

where

ΔQ = change in radiative heating of the surface–troposphere system

$\Delta Q(A)$ = change in radiative heating of troposphere

$\Delta Q(s)$ = change in radiative heating of the surface

$\Delta F{\downarrow}(O)$ = change in downward F at the surface

$\Delta F\!\uparrow(h)$ = change in upward F at the tropopause
$\Delta F\!\downarrow(h)$ = change in downward F at the tropopause.

Trace gases absorb and emit radiation as functions of wavelength in discrete lines with extended wings. Dickinson and Cicerone (1986) have pointed out that the relationship between radiative forcing change at the tropopause and concentration change is approximately logarithmic for large concentrations such as CO_2, square root at intermediate values such as CH_4 and N_2O, and linear at low concentrations such as CFCs. Using the results of the radiative transfer calculations for CO_2 and CH_4 done by Kiehl and Dickinson (1987) with a 1-d narrow-band radiative convective model, and those for N_2O and the CFCs done with a 1-d radiative convective model, as reported by Ramanathan et al. (1985), Wigley (1987) derived the following formulas for calculating a change in radiative forcing (ΔQ) resulting from a concentration change from C_0 to C:

<table>
<tr><td></td><td></td><td>Valid over
Concentration Range</td></tr>
<tr><td>CO_2:</td><td>$\Delta Q = 6.333 \ln (C/C_0)$</td><td>250–600</td></tr>
<tr><td>CH_4:</td><td>$\Delta Q = 0.0395 (\sqrt{C} - \sqrt{C_0})$</td><td>700–3100</td></tr>
<tr><td>N_2O:</td><td>$\Delta Q = 0.105 (\sqrt{C} - \sqrt{C_0})$</td><td>285–560</td></tr>
<tr><td>CFC 11:</td><td>$\Delta Q = 0.27\, C(\text{CFC } 11)$</td><td>0.1–2.0</td></tr>
<tr><td>CFC 12:</td><td>$\Delta Q = 0.31\, C(\text{CFC } 12)$</td><td>0.2–3.5</td></tr>
<tr><td>All CFCs:</td><td>$\Delta Q = 1.052\, C(\text{CFC } 11)$</td><td></td></tr>
</table>

$$\Delta Q(\text{total}) = 6.333 \ln ([CO_2]/C_0[CO_2])$$
$$+ 0.0398 (C[CH_4]^{0.5} - C_0[CH_4]^{0.5})$$
$$+ 0.105 (C[N_2O]^{0.5} - C_0[N_2O]^{0.5})$$
$$+ 1.052\, C(\text{CFC } 11)$$

C and C_0 are in ppm for CO_2, and in ppb for all other gases; ΔQ is in Wm^{-2}.

The models on which these expressions are based account for CH_4–CO_2–H_2O overlap, assuming that other overlap effects are likely to be small. It should be stressed here that the formulas are only valid over the indicated concentration range for each gas.

This leads to another important question, namely whether the changes in tropospheric radiative forcing due to greenhouse gases are additive. Taking into account the various trace-gas band overlaps, model calculations indicate an error of about 10 percent due to the assumption of additiveness (Ramanathan et al. 1987). This finding may, however, not be valid if one or more of the radiatively active gases are altered through chemical interactions. For example, the chemistry involving CH_4, CFCs, and O_3 is quite different in the various atmospheric layers, thereby producing different changes in the vertical O_3 profile. Since the vertical O_3 distribution changes have an influence on the tropospheric effects, the assumption of the validity of additiveness needs further study.

Finally, the global average equilibrium surface temperature (ΔT_e) is related to the change in radiative forcing (ΔQ) by the following simple expression:

$$\Delta T_e = \lambda \Delta Q$$

where

$$\lambda = \frac{F}{\left[\dfrac{dF}{dT_e} + \dfrac{S}{4} \dfrac{d\alpha_p}{dT_e} \right]}$$

λ = climate sensitivity parameter

$1/\lambda$ = global climate feedback parameter, which is the total of the negative and positive feedbacks (Dickinson 1986)

F = longwave radiative flux

S = solar constant

α_p = planetary albedo.

The λ values from a variety of model calculations as shown by Ramanathan et al. (1985) range from 0.47 to 0.53 K/Wm^{-2}. In our calculations, we have used their latest estimate of 0.52.

For a CO_2 doubling, we get

$$\Delta Q(2 \times CO_2) = 6.333 \ln 2$$
$$= 4.39 \text{ Wm}^{-2}$$
$$\lambda = 0.52 \text{ K/Wm}^{-2}$$
$$\lambda T_e(2 \times CO_2) = 0.52 \text{ K/Wm}^{-2} \times 4.39 \text{ Wm}^{-2}$$
$$= 2.28 \text{ K}$$

This means that our temperature-change estimates presented in the main body of this book have a climate sensitivity of 2.3 K. This compares with the climate sensitivity of 1.5–4.5 K obtained by the GCMs, which are the most sophisticated climate models.

REFERENCES

Dickinson, R. E. (1986), in B. Bolin, R. Doos, J. Jaeger, and R. A. Warrick, eds., *The Greenhouse Effect: Climatic Change and Ecosystems*, Wiley, New York, pp. 207–270.

Dickinson, R. E., and R. Cicerone (1986), "The Climate System and Modeling of Future Climate," in B. Bolin, R. Doos, J. Jaeger, and R. A. Warrick, eds., *The Greenhouse Effect: Climatic Change and Ecosystems*, Wiley, New York, pp. 206–270.

Kiehl, J., and R. E. Dickinson (1987), "A Study of the Radiative Effects of Enhanced Atmospheric CO_2 and CH_4 on Early Earth Surface Temperatures," *Journal of Geophysical Research*, Vol. 92, D3, pp. 2991–2998.

Kiehl, J., and Ramanathan, V. (1982), "Radiative Leaking due to Increased CO_2: The Role of H_2O Continuum Absorption," *Journal of Atmospheric Science*, Vol. 39, pp. 2923–2926.

Ramanathan, V., et al. (1985), "Trace Gas Trends and Their Potential Role in Climate Change," *Journal of Geophysical Research*, Vol. 90, pp. 5547–5566.

Ramanathan, V., et al. (1987), "Climate-Chemical Interactions and Effects of Changing Atmospheric Trace Gases," *Reviews of Geophysics*, Vol. 25, No. 7, pp. 1441–1482.

Wigley, T. M. L. (1987), "Relative Contributions of Different Trace Gases to the Greenhouse Effect," *Climate Monitor*, Vol. 6, No. 1, pp. 14–28. '

Index